Electricity

10th Edition

by

Howard H. Gerrish

William E. Dugger, Jr.
Director
Technology for All
Americans Project

Kenneth P. DeLucca
Professor, Department of
Industry and Technology
Millersville University
Millersville, Pennsylvania

Publisher
The Goodheart-Willcox Company, Inc.
Tinley Park, Illinois
www.g-w.com

Copyright © 2009

by

The Goodheart-Willcox Company, Inc.

Previous editions copyright 2004, 2001, 1994, 1988, 1978,
1975, 1968, 1966, 1964

All rights reserved. No part of this work may be reproduced, stored, or transmitted in any form or by any electronic or mechanical means, including information storage and retrieval systems, without the prior written permission of The Goodheart-Willcox Company, Inc.

Manufactured in the United States of America.

Library of Congress Catalog Card Number 2008042977

ISBN 978-1-60525-041-0

1 2 3 4 5 6 7 8 9 – 09 – 13 12 11 10 09

The Goodheart-Willcox Company, Inc. Brand Disclaimer: Brand names, company names, and illustrations for products and services included in this text are provided for educational purposes only and do not represent or imply endorsement or recommendation by the author or the publisher.

The Goodheart-Willcox Company, Inc. Safety Notice: The reader is expressly advised to carefully read, understand, and apply all safety precautions and warnings described in this book or that might also be indicated in undertaking the activities and exercises described herein to minimize risk of personal injury or injury to others. Common sense and good judgment should also be exercised and applied to help avoid all potential hazards. The reader should always refer to the appropriate manufacturer's technical information, directions, and recommendations; then proceed with care to follow specific equipment operating instructions. The reader should understand these notices and cautions are not exhaustive.

The publisher makes no warranty or representation whatsoever, either expressed or implied, including but not limited to equipment, procedures, and applications described or referred to herein, their quality, performance, merchantability, or fitness for a particular purpose. The publisher assumes no responsibility for any changes, errors, or omissions in this book. The publisher specifically disclaims any liability whatsoever, including any direct, indirect, incidental, consequential, special, or exemplary damages resulting, in whole or in part, from the reader's use or reliance upon the information, instructions, procedures, warnings, cautions, applications, or other matter contained in this book. The publisher assumes no responsibility for the activities of the reader.

Library of Congress Cataloging-in-Publication Data

Gerrish, Howard H.
  Electricity / by Howard H. Gerrish, William E. Dugger, Kenneth P DeLucca. — 10th ed.
    p. cm.
  Includes index.
  ISBN 978-1-60525-041-0
  1. Electric engineering. I. Dugger, William. II. DeLucca, Kenneth P. III. Title.
TK146.G43 2009
537—dc22   2008042977

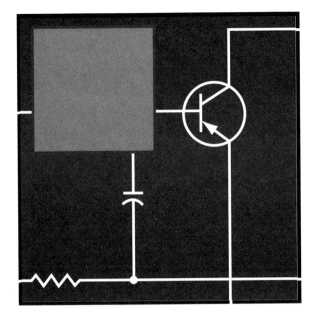

Introduction

A career in the expanding field of electricity and electronics offers young people an opportunity for satisfying and productive employment. This exploratory course is intended to familiarize all students with basic principles of electricity and their practical applications.

The course emphasizes how many ways electricity affects our everyday living. It helps students acquire safe work habits and develop skills in using tools in simple electrical construction.

These instructional units provide, in easy-to-understand language, information on sources of electricity, types of circuits, operating principles of electric motors, measuring instruments, generators and transformers, inductance, capacitance, and an introduction to solid-state devices.

It is the desire of the authors that this text will contribute much toward the general educational development, career preparation, and enjoyment to all who use this text.

To the Student

Electricity is a fascinating subject to study. As you progress through the basic principles on which knowledge of the electrical phenomena is based, you will realize the important role electricity plays in our everyday lives. It is one of the greatest forces we have harnessed.

You will find that the simple projects you build in connection with your studies will demonstrate and give understanding to the laws and applications of electricity. Your interest will be stimulated. Electricity is an exciting science.

The study of electricity is not only for the scientist. This force has molded our civilization and raised our standard of living. Having an understanding of its principles and application is a basic requirement of any well-educated person. Electricity offers you a challenge. Will you accept it?

Howard H. Gerrish
William E. Dugger, Jr.
Kenneth P. DeLucca

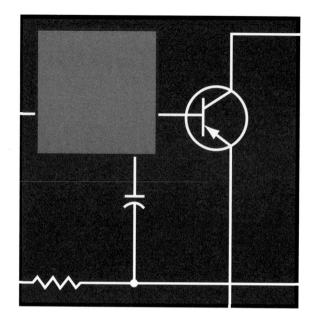

Table of Contents

Chapter 1
The Electron 9
  Electricity—An Introduction
  Classification of Elements
  Ionization
  Conductors and Insulators
  Law of Charges
  Wire Sizes

Chapter 2
Volts, Amperes, Ohms ...15
  The Basics of Any Electrical Circuit
  Resistance and Conductor Size
  Conductance
  Stranded Conductors
  Superconductors
  Uses for Resistance
  Resistor Color Code

Chapter 3
Meters, Reading a Meter ..23
  Analog/Digital Meters
  d'Arsonval Meter
  Iron-Vane Meter
  Ammeters and Meter Shunts
  Voltmeters and Multipliers
  Ohmmeters
  Meter Connections
  Learning to Read a Meter
  Analog/Digital Meters—Which One
    Is Best?
  Meter Precautions

Chapter 4
Ohm's Law 35
  The Simple Circuit
  Ohm's Law
  Overload Protection of Circuits
  Circuits and Switches

Chapter 5
Power 43
  Power
  Horsepower
  Earth and the Environment: How to
    Conserve Electrical Power
  Electric Motors

Table of Contents

Chapter 6
Series Circuits 49
    Voltage Drop
    Summary of the Laws of a Series Circuit

Chapter 7
Parallel Circuits 55
    Equal Resistors in Parallel
    Unequal Resistors in Parallel
    Conductance
    Two or More Resistors in Parallel
    Equivalent Resistance
    Applications

Chapter 8
Sources of Electricity–Batteries 65
    Sources of Electricity
    The Voltaic Cell
    Cell Connections
    Primary Cells
    Secondary Cells
    Series and Parallel Connections
    Battery Capacity

Chapter 9
Sources of Electricity–Friction, Heat, Pressure, Light 73
    Voltage from Friction
    Voltage from Heat
    Voltage from Pressure
    Voltage from Light

Chapter 10
Magnetism 77
    Lodestones
    Permanent Magnets
    Laws of Magnetism
    Magnetic Field Pictures
    Electricity and Magnetism
    Electromagnets
    Relays
    Solenoids as Switches
    Hall Effect Devices
    Magnetism by Induction
    Magnetic Shielding

Chapter 11
Motors 87
    How the Motor Works
    The Commutator
    Motor Circuits
    The Complete Motor
    Motor Principles

Chapter 12
Direct Current Generators 95
    Direction of Current
    The Generator
    The Commutator
    Generator Output
    Generator Losses
    Field Excitation

Chapter 13
Alternating Current ...103
    Vectors and Sinusoidal Waveforms
    Frequency
    Period
    Average Values
    Effective Values
    Peak-to-Peak Values

Chapter 14
Capacitance 111
Capacitors
The Capacitive Circuit
The Farad
Working Voltage
Labeling Capacitors
Factors that Determine Capacitance
Kinds of Capacitors
RC Time Constants
Capacitive Reactance
Impedance

Chapter 15
Inductance 121
Self Inductance
The Henry
Labeling Inductors
Coil Action in a DC Circuit
Inductive Reactance
Impedance
Combining X_L and X_C in a Circuit
Resonance

Chapter 16
Transformers 129
Lenz's Law
Transformer Losses
Transformer Construction
Turns Ratio
Power Transmission
The Induction Coil
Autotransformer

Chapter 17
Semiconductors 139
Amplification
Semiconductor Current
Diodes
Light-Emitting Diodes
Rectification
Transistors
Transistor Circuits
Transistor Amplifiers and Alpha
Common Emitter Circuit and Beta
The Transistor Switch
Other Special Semiconductor Devices

Chapter 18
Integrated Circuits 153
Basing Diagrams for ICs
Digital and Linear Integrated Circuits
Computers

Chapter 19
Electrical/Electronic
Projects 161
Soldering Primer
Desoldering Primer
Project 1 Experimenter
Project 2 Audio Oscillator
Project 3 Transistor Radio Receiver
Project 4 Door Chime
Project 5 Magnetic Relay
Project 6 Induction Coil
Project 7 Two-Pole Motor
Project 8 Electric Engine
Project 9 Continuity Tester
Project 10 Automotive Battery Charger

Glossary 195

Index 203

Electricity and Electronics Lab Safety

All the rules of general safety used in other school laboratories apply equally to the electrical/electronics lab. So that you are aware of these, some of the more important ones will be repeated.

1. Many painful accidents occur by careless and thoughtless actions—think!
2. Your instructor is there to help. Ask for his or her approval before starting activities. This will save time and may help prevent accidents.
3. Report any injury at once. A small cut can develop into serious trouble if not properly cared for.
4. Your eyes are a priceless possession. Wear safety goggles when grinding or when working where sparks or chips are flying. It is also wise to wear eye protection when soldering or when working around automotive (lead-acid) storage batteries.
5. Keep the lab floors clean and free of litter that might cause someone to slip or stumble.
6. Use tools correctly and do not use them if they are not in proper working condition.
7. Observe proper ways of handling and lifting objects. Get help to lift heavy objects.
8. Do not talk to or attract the attention of anyone else when operating machinery.
9. Never leave a machine running or coasting/running down to stop. Wait for it to stop completely before leaving it.
10. Obtain permission from your instructor before using any power tools.

Special Safety Rules for the Electricity/Electronics Lab

1. Observe safety rules concerning each project and be particularly careful not to contact any wire or terminal that is connected to a high voltage.
2. When testing your projects, always keep one hand in your pocket. If two hands are in contact with a circuit, a current could potentially flow through your body causing death or serious injury.
3. Do not apply voltage or turn on any device until it has been properly checked by the instructor.
4. A project turned off and disconnected from the power source can still contain a charge of electricity. Always short out capacitors with a low value, high wattage resistor or with an insulated screwdriver before attempting to work on or repair a project involving a capacitor.
5. If your project should blow a fuse in the main power line, do not turn it on until the trouble is discovered and remedied. Have your instructor assist you.
6. Always stand at a safe distance and look away from any project when it is turned on for the first time to minimize possible harm if there is a problem with the circuit. Be sure your instructor is present.
7. Know where the fire extinguishers are placed in your laboratory.
8. Wires carrying excessive current, resistors and vacuum tubes get hot while operating. Wait for them to cool before attempting to touch them.
9. Be sure equipment is in proper working order before use. Frayed cords and damaged plugs are a major source of accidents.
10. Ask for instruction before using any piece of electronic test equipment. One wrong connection can destroy an instrument. Repair or replacement of electronic instruments can be expensive.

Special Safety Rules for Soldering

1. Always think that a soldering iron is HOT! Be careful and try to prevent burns.
2. Do not solder near materials that can readily burn.
3. Replace the soldering iron in its holder or tray when not in use.
4. Remember, a hot soldering iron can melt electrical cords, providing for an unsafe electrical situation.
5. Before you attempt to change soldering iron tips, make sure the soldering iron has fully cooled. It is also a good idea to unplug the soldering iron.
6. Always solder in a well-ventilated area.
7. Always wear safety glasses to prevent hot solder from splashing into your eyes.
8. Remember, lead is a very toxic material. Do not eat, drink, or put hands or fingers in your mouth or near your eyes after handling solder. Always wash your hands after handling solder. There are special soaps to assist in the removal of lead from your hands after soldering. These cleaning agents can be obtained at some specialty stores and stores that provide materials for stained glass window projects.

Safety Quiz

1. Why does an electronics technician make tests on a live circuit with one hand in a pocket?
2. Describe the times when safety goggles should be worn in the electricity/electronics laboratory.
3. Why should you stand away from and look away from your project when it is turned on for the first time?
4. Which parts of a circuit might become hot during operation?
5. Why is a project still dangerous, even after the power is turned off?
6. If a device blows a fuse, should you replace the fuse and again operate the device? Explain.
7. Why should you wash your hands with special soap after handling solder?

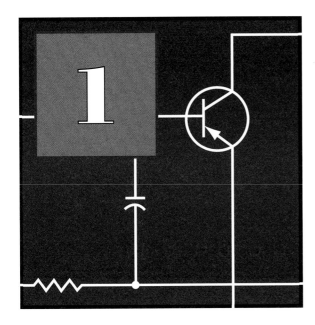

The Electron

Objectives

After studying this chapter, you will be able to answer these questions:
1. What is the nature of matter?
2. What is electron theory?
3. How does an electric current flow through a wire or conductor?

Important Words and Terms

The following words and terms are key concepts in this chapter. Look for them as you read this chapter.

American Wire Gauge	insulator
atom	ionized
atomic number	mass number
atomic weight	mil
circular mil	molecule
conductor	neutral
coulomb	neutron
current	nucleus
electricity	proton
electron	resistance
electrostatic field	semiconductor

Electricity—An Introduction

Spectacular discoveries and inventions have been realized in the science of electricity and electronics. The use of electricity has become such a common part of our everyday life that one seldom thinks about the vast wired and wireless networks that make it possible for us to use the great invisible force called electricity.

You turn on your CD or MP3 player. You use the remote control to turn on the television to play your favorite video game. You snap a switch on the wall and immediately a room is filled with light. Electricity is commonly used for refrigeration, cooking, and for washing and drying clothes. It is used to facilitate heating, cooling, mixing food, kitchen ventilation, garbage disposal, and for a multitude of other uses. Electricity powers the battery chargers for our cell phones and electronic games. Electricity is used to power the computers that allow us to communicate with the world!

Electricity, and its use in electronic applications, made possible the first walk on the moon. Current space vehicle flybys to the furthest reaches of our solar system are also taking place thanks to modern electronics.

Whatever a person's chosen career path may be, a fundamental knowledge of electricity and electronic principles should be a part of his or her general education.

Electricity is not new. It has been in existence since the beginning of time. Only in recent years have scientists explored the phenomena of electricity and proposed theories as to its nature. It seems strange that we do so many things with electricity and yet no scientist has ever seen it. One might call it "the great invisible wonder of the twentieth century," because this was the time that electricity was "put to work!"

Over two thousand years ago the Greeks discovered that if a yellowish brown translucent resin called amber was rubbed very hard with a cloth, it would attract small pieces of dust and dried grass. They were seeing static electricity in action. The Greeks believed that these amber fossils were living stones. They called them "elektron." From this Greek name is derived "electronics."

As we are interested in the science of electronics, a more thorough understanding of the electron is necessary.

All matter of substance is made up of molecules. Let's see just what that means. If you could take a glass of water and keep dividing it and dividing it, you would finally reach a point at which no further division could be made and still keep the identity of the water. We could say that a *molecule* is the smallest division of a substance that could be made without destroying the identity or properties of the substance.

You know that water is a chemical combination of hydrogen and oxygen. We refer to it as H_2O. Neither hydrogen nor oxygen alone has any resemblance to water at all. They are entirely different, but when united, molecules of water are formed. These particles of hydrogen and oxygen, which make up the molecule of water, are known as atoms.

The *atom* is the smallest particle of an element. There have been over a hundred different elements discovered. The chemist arranges them in order of their weights, or groups them according to similar properties. This arrangement is called the *Periodic Table of the Elements*. If you go to any chemistry classroom, this table will certainly be posted.

You are familiar with elements such as copper, silver, and iron. There are tin, lead, gold, uranium, as well as many others. The Greeks thought that all matter was made up of these atoms, and that the atom was the smallest particle that could exist. This theory was advanced by the English physicist John Dalton in 1808.

The atom has been "smashed" to study it further. Physicists are exploring the

History Hit!

John Dalton (1766-1844)

Dalton's interest was in weather and chemistry. The son of a weaver, he grew up in the isolated village of Cumbria, England. He worked as a teacher, and in 1808 he determined that every element consists of very small particles called atoms, indivisible and indestructible spheres. Dalton's theory on atoms has been amended and refined by others in the following years. However, his theory's overall structure still serves as the central basis of modern chemistry and physics.

structure of the atom itself. They discovered that the atom contains a center called the *nucleus.* This nucleus, which contains most of the mass of the atom, is made up of a certain number of positive particles called *protons* and a number of neutral particles called *neutrons.* Revolving around this core or nucleus, in orbits, are negatively charged particles called *electrons.* This is shown in **Figure 1-1.**

Classification of Elements

Elements are classified in the Periodic Table of the Elements by the number of protons in the nucleus. This is the *atomic number.* Also, elements are classified by the number of protons and neutrons in the nucleus. This is the *atomic weight* or *mass number.* The atomic models shown in **Figure 1-2** will help you to understand this.

Ionization

Most often, all atoms have the same number of electrons in orbit as the number of protons in the nucleus. Therefore, the atom is electrically in balance or *neutral.*

Some atoms hold their revolving electrons rather loosely. This loose hold allows the transfer of electrons between atoms. When electrons transfer, atoms will gain or lose electrons. When this occurs, the atom becomes unbalanced with either an excess or a deficiency of electrons. This

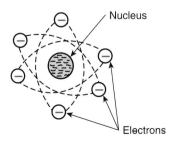

Figure 1-1.
Electrons in orbit around the nucleus.

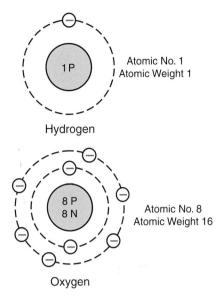

Figure 1-2.
Atomic models of hydrogen and oxygen.

atom is said to be *ionized.* If the atom loses some electrons, it has more protons than electrons and is positively charged. It is a positive ion. On the other hand, if the atom gains some electrons, it has more electrons than protons and is negatively charged. It is a negative ion.

Conductors and Insulators

Some elements, particularly the metallic elements such as copper, silver, aluminum, and others, hold their electrons rather loosely. A very small force will cause them to give up some electrons. Such an element is a good conductor of electricity. Electrical energy is transferred through the conductor by the free movement of electrons from one atom to the next atom as shown in **Figure 1-3.**

As an electron is added to one end of the conductor, an electron leaves the other end of the conductor. This transfer of electrons through the conductor is called *current* or *electricity.* It is important to observe carefully that the actual electrons move only short distances as they displace

12   Electricity

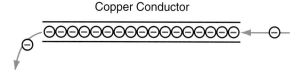

Figure 1-3.
The transfer of energy by electron movement is called electric current.

each other, but the actual transfer of energy from one end of the conductor is almost instantaneous.

The speed of this transfer has been measured and found to be near 186,000 miles per second or 300,000,000 meters per second. In other words, electricity can transfer its energy at the speed of light!

A material that allows the free movement of many electrons is a good *conductor* of electricity. On the other hand, if a material allows only a few electrons to move freely, it is considered an *insulator.* *Semiconductors* are elements that are neither true conductors nor insulators in their ability to permit electrons to flow. Opposition to the flow of electrons created by the material is called *resistance.* A good conductor has a low resistance. An insulator has a high resistance, and semiconductors lie somewhere in between. Some familiar materials of each type are listed in the table in **Figure 1-4**.

Law of Charges

One of the most important lessons in the study of electrons and electrically charged particles is the attraction and repulsion of differently charged particles. The law of charges states that *like charges repel and unlike charges attract.* You *must* understand this thoroughly. This principle will be used many times in future lessons to explain the concept of electricity.

Existing in space around a charged body is an invisible field of force called an *electrostatic field.* This phenomenon is easily demonstrated by experiments conducted in the laboratory. These static fields of force and their direction of force are represented by the small arrows in **Figure 1-5** and **Figure 1-6**. Figure 1-5 shows the case of two negatively charged bodies. Note the fields repel each other and cause the bodies to move away from each other. In Figure 1-6, the fields are attracted to each other and create a force that causes the bodies to move toward each other.

History Hit!

Charles Coulomb (1736-1806)

A French physicist, Coulomb discovered the relationship between electrical and magnetic attraction and repulsion. He trained and served as a military engineer. The basic unit of electrical charge is named in his honor, the coulomb.

Conductors	Semiconductors	Insulators
Silver	Silicon	Air
Copper	Germanium	Glass
Aluminum	Selenium	Porcelain
Brass	Carbon	Rubber
Iron		Bakelite

Figure 1-4.
Examples of materials that can be used as conductors, semiconductors, and insulators.

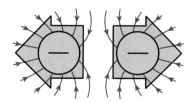

Figure 1-5.
Like charges repel each other due to opposing electrostatic fields. Repulsion is indicated by large arrows.

Chapter 1   The Electron

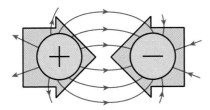

Figure 1-6.
Unlike charges attract each other due to attractive electrostatic fields. Attraction is indicated by large arrows.

AWG Size	D (Approx.) in Mils	Cross Section Area Circular Mils
8	128.0	16,500
10	102.0	10,400
12	81.0	6,530
14	64.0	4,110
16	51.0	2,580
18	40.0	1,620
20	32.0	1,020
22	25.3	642
24	20.1	404
26	15.9	254

Figure 1-7.
Sizes of wires commonly used for electrical circuits.

The force of the attraction and repulsion of charged particles was explored by the famous scientist, Charles Coulomb. To describe the difference in charge between two bodies, it would be insignificant to say that this body has one or two more electrons than the other, because an electron is such a tiny particle of electricity. A more practical unit of measurement is a much larger number of electrons. One such measurement is the coulomb. A *coulomb* represents 6,240,000,000,000,000,000 electrons or 6.24×10^{18} electrons.

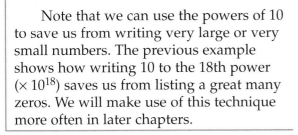

Note that we can use the powers of 10 to save us from writing very large or very small numbers. The previous example shows how writing 10 to the 18th power ($\times 10^{18}$) saves us from listing a great many zeros. We will make use of this technique more often in later chapters.

Wire Sizes

Conductors or wires for electrical circuits are manufactured in many sizes, materials, and with many types of insulation coverings. The number assigned to each size is known as the *American Wire Gauge (AWG).* For example, a commonly used wire size in the wiring of a house is AWG No. 12. As the number becomes larger, actual wire size decreases. Likewise, as the number becomes smaller, wire size increases.

In the table shown in **Figure 1-7,** several common sizes are listed. Also listed are their diameters in *mils* (one thousandth of an inch) and their cross-sectional area in circular mils. A *circular mil* is the cross-sectional area of a wire 0.001 inches in diameter. The area of a conductor in circular mils is found by squaring the diameter of the conductor measured in mils.

For example: A wire has a diameter of 50 mils, what is its area in circular mils?

$$50^2 = 2500 \text{ circular mils}$$

Do not be confused by the term circular mil. It is different than a square mil and is a smaller area. The use of the circular mil saves a great deal of time doing mathematics to compute sizes of wires.

A standard American Wire Gauge is shown in **Figure 1-8.** You will want to measure an assortment of wires with this gauge. While learning the AWG system, keep in mind that metric measurements are being used more and more for wire sizes. As American companies trade with other countries of the world, those countries are requiring the use of metric equivalents or metric sizes. For example, AWG No. 12

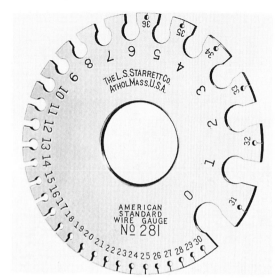

Figure 1-8.
A gauge used to determine wire size.
(L.S. Starrett Co.)

wire has a standard metric size of 2.0 millimeters (mm). AWG No. 10 wire (smaller number, thicker wire) has a diameter of 2.5 mm.

Web Wanderings!

http://www.nobelprize.org/

Visit the Nobel Prize Web site for articles, interviews, biographies, and illustrated presentations on Nobel Prize Laureates and their accomplishments. Click on the *Nobel Prizes* link. By following the Physics link, you'll find detailed and entertaining information on physics- and electronics-related topics. There, you can play educational games, view multimedia presentations, and read articles on subjects such as matter, energy, semiconductors, and vacuum tubes. Note: Web sites do change their addresses often. You may need to search for sites and links carefully if listings are revised and updated.

Quiz–Chapter 1

Write your answers to these questions on a separate sheet of paper. Do *not* write in this book.

1. Like charges _____ each other.
2. Unlike charges _____ each other.
3. The word electronics is derived from the Greek word _____.
4. The smallest division of a compound that can be made without the compound losing its identity is known as a(n) _____.
5. Compounds are made up of _____ and were once considered the smallest particle that could exist.
6. Elements are arranged in a table by their atomic _____ and their atomic _____.
7. List four good conductors of electricity.
8. List four good insulators.
9. When an atom has lost or gained some electrons, it is said to be _____.
10. Opposition to the flow of electrons in a substance is known as _____.
11. A neutral part of an atom is called a(n) _____.
12. A positively charged part of an atom is called a(n) _____.
13. A negatively charged part of an atom is called a(n) _____.
14. A material that has resistance in between that of a conductor and an insulator is a(n) _____.

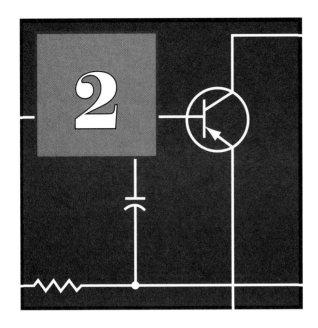

Volts, Amperes, Ohms

Objectives

After studying this chapter, you will be able to answer these questions:
1. What is the unit of electromotive force called a volt?
2. How is an electrical current measured?
3. How is resistance to the flow of current measured?

Important Words and Terms

The following words and terms are key concepts in this chapter. Look for them as you read this chapter.

amp	potential difference
ampere (A)	potentiometer
color code	rheostat
conventional current flow	superconductor
	tolerance
electromotive force (emf)	voltage (V or E)
electron flow	volts (V)
ohm (Ω)	

The Basics of Any Electrical Circuit

Each electrical circuit will contain these basic electrical quantities: *voltage, amperage,* and *resistance.* Each of these quantities must be controlled to permit the electrical circuit to operate properly. Each of these terms will be defined and examined to understand just how they apply in any electrical circuit.

Volts

An interesting experiment performed in the laboratory is shown in **Figure 2-1.** Container A is connected to container B by a pipe. When A is filled with water, the water flows through the pipe to B until the level of water in both containers is the same.

If we should take two terminals and cause one to have an excess of electrons (negative) and the other to have fewer electrons (positive), then connect a conductor (pipe) between them, electrons will flow from one to the other until the number of electrons on one becomes equal to the electrons on the other, **Figure 2-2.**

16 Electricity

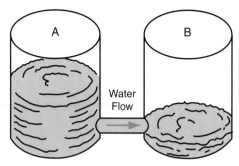

Figure 2-1.
Water will flow until the levels in the two containers become equal.

What caused the water to flow? It was the difference between the level of the water in each container. What caused the water to stop flowing? The level of the water in each container became the same or equal.

What caused electrons to flow between the two oppositely charged bodies? It was the difference between the number of electrons. In electricity, this is called the *potential difference.* What caused the electrons to stop flowing? The number of electrons became equal and the potential difference became zero. It is easy to understand that electrons flowed because of a potential difference that created a force, called an *electromotive force (emf)* or *voltage (V or E).* This force existed only during the time that the electrons were unevenly distributed between the two terminals.

In the case of the water, the force that caused the water to flow can be measured in pounds per square inch. The greater the difference between the two levels, the greater the force.

In the case of the charged terminals, the force that causes the electrons or current to flow (the voltage or electromotive force) is measured in *volts (V).* The letter symbol for voltage is the letter E or V. The actual value of one volt is accurately kept by the National Institute of Standards and Technology (NIST). Formerly NIST was known as the National Bureau of Standards.

Amperes

If you wished to do so, you might measure the number of gallons of water per minute flowing through a garden hose. You would only need to see how many gallon cans were filled in one minute. The flow of water through the hose could be considered as the quantity of water that passed any given point in the hose, and the flow is the same at any point at which you might measure, **Figure 2-3.**

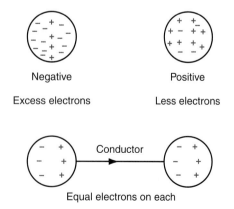

Figure 2-2.
Unequal electrical charges will cause electrons to flow.

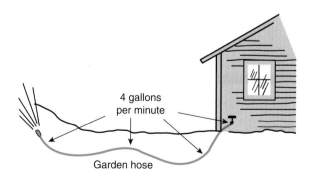

Figure 2-3.
The flow of water in the hose is the same at any point.

Chapter 2   Volts, Amperes, Ohms

History Hit!

André-Marie Ampère (1775-1836)

Ampère was a French physicist and mathematician. He was largely self-taught. He had a photographic memory and loved reading. Ampère was a university professor, and he was appointed by Napoleon as the inspector-general of the entire university system of France. He enjoyed physics, philosophy, psychology, and chemistry. His most well known contribution was concerning electrical currents and magnetism. For this reason, he is honored with his name being attached to the measurement of electrical current—the ampere (sometimes call the amp, which should not be confused with the word amplifier).

In the case of electric current, flow can be measured by a definite quantity of electrons passing any given point in the conductor. We learned in Chapter 1, that the unit of a given quantity of electrons is known as the *coulomb*, and represents a very large number of electrons. If one coulomb of electrons passes a fixed point in a conductor per second, one *ampere (A)*, or *amp*, of current is said to be flowing. This unit of measurement was named in honor of the French scientist, André-Marie Ampère. The symbol for current is the letter *I*.

Some textbooks on electricity describe the flow of electricity as flowing from positive to negative. This is called *conventional current flow*. In this text, *current* will always be assumed to be *electron flow* and will flow from negative to positive. See **Figure 2-2**. Realize that conventional current flow and electron flow are both methods used to explain the operation of electricity. One method is not more correct or better than the other. Just as English is spoken in America while German is spoken in Germany. Each technique or language is acceptable in that area. In our study of electricity and electronics, this will be an important fact to remember.

Ohms

If your garden hose has a diameter of five-eighths inches (the hose is five-eighths of an inch thick), you can see that there are limits to the amount of water the hose can carry. If you need more water, you can buy a larger hose or even use two hoses. In other words, the size and internal friction of the hose will limit the quantity of water that flows through it, without damage to the hose.

The same principle applies to an electrical conductor. A certain size and kind of wire offers a definite opposition or limitation or *resistance* to the flow of an electric current. If large currents are forced through the conductor that exceed its ability to carry current, the wire will become very hot and it could melt and be destroyed. Many homes burn down each year because the wire used is inadequate in size to carry the electrical current necessary for today's modern living.

In a properly wired home, there are either fuses or circuit breakers installed at the electrical service entrance panel (sometimes called the breaker box). These are safety devices. A fuse will burn out or a circuit breaker will open if the circuit is forced to carry too much current.

The resistance to the flow of current is measured in *ohms* (Ω). One ohm of resistance will allow one ampere to pass when one volt pressure is applied.

The resistance of a conductor depends upon:
1. The *length* of the conductor. If a wire has a resistance of one ohm per hundred feet, then 1000 ft. would have a resistance of 1×10 or 10 ohms.
2. The *size* of the conductor. The larger or thicker the wire, the less resistance.

3. The *material* used for the conductor. Materials differ in their ability to conduct electricity. For instance, copper is a better conductor than aluminum. Gold and silver are among the best conductors.
4. The *temperature* of the material. In common conductors such as copper and aluminum, the resistance increases as the temperature increases.

Resistance and Conductor Size

If the power company is going to conduct electric power to cities, homes, and factories, the size and kind of wires used become an important consideration. The material from which a conductor is made must have a low resistance and be of sufficient size to carry the electric current. It must also have the physical strength to withstand the rigors of snow, ice, and wind.

Silver is the best conductor but it is very expensive. Copper is also an excellent conductor. It has high tensile strength (meaning it can be pulled at each end without readily breaking), but it is heavy and also quite expensive. Aluminum offers some distinct advantages for use in high voltage transmission lines. Aluminum is light and relatively inexpensive. Because of its light weight, it is possible to run long spans of aluminum wire between supports and thus reduce the number of supports needed. However, aluminum has more resistance to electrical current. Aluminum conducts electricity about sixty percent as well as copper.

An example of resistance and wire sizes used in the home can be helpful. The lights in your home can satisfactorily operate with wire size AWG No. 14 (1.6 mm). An electric range may require AWG No. 6 (4.0 mm) wire or larger to feed the necessary current to the range.

Figure 2-4 lists several sizes of copper wire and the resistance per 1000 feet of wire. Study this table and notice how the resistance increases as the wire grows smaller in thickness (diameter).

Math Manipulation!

Problem: What is the resistance of 300 feet of AWG No. 16 (1.25 mm) solid copper wire?

Explanation: Referring to the table, AWG No. 16 (1.25 mm) wire has a resistance of 4.73 ohms per thousand feet. Therefore, the resistance of 300 feet would be:

$$\frac{300}{1000} \times 4.73 = 0.3 \times 4.73 = 1.419 \text{ ohms}$$

Another example will show a different use of this table.

Problem: A wire must be run over a distance of 500 feet and the resistance cannot exceed one ohm. What size wire should be used?

Explanation: As 500 feet is one half of one thousand, it is only necessary to look at the table and discover which size wire has a resistance of less than two ohms per 1000 feet. You would select AWG No. 12 (2.00 mm) because:

$$0.5 \times 1.87 = 0.935 \text{ ohms}$$

Conductance

Up to this point we have only considered copper and aluminum as conductors. You should understand that as the resistance of a wire decreases, its ability to conduct increases and vice versa. So you could state the current carrying capacity of a wire either by stating its resistance as measured in ohms or its ability to conduct.

Chapter 2   Volts, Amperes, Ohms

Wire Size AWG	Metric Size Millimeters (mm)	Resistance Ohms/1000 Ft.
6	4.00	0.465
8	3.15	0.739
10	2.50	1.18
12	2.00	1.87
14	1.60	2.97
16	1.25	4.73
18	1.00	7.51
20	0.80	11.9
22	0.63	19.0
24	0.5	30.2

Figure 2-4.
Table showing resistance offered by several sizes of solid copper wire.

Conductance will be studied in a later chapter, when computing parallel circuits.

Referring to **Figure 2-5,** you will observe the relative conductivity of several common materials. This table is based on the fact that silver is the best conductor and therefore is assigned the value 100. All other materials are less, based on a percentage compared to silver.

It is apparent from Figure 2-5 why carbon is used as a resistive material in electric and electronic circuits. It has a very low ability to conduct.

Stranded Conductors

There are occasions when one solid wire is not satisfactory for a particular use. One case would be for use in a wire that will receive a lot of bending such as an appliance cord. If made of one solid wire, it would be stiff rather than flexible and

Material	Conductivity
Silver	100
Copper	98
Aluminum	61
Tungsten	32
Iron	16
Carbon	.05

Figure 2-5.
Comparing conductivity of several materials.

would break after a few bends. It is more practical to make up a cord of several strands of smaller wire.

Another case is where a group of conductors is used instead of the solid conductor in very large power transmission lines, as described previously. It is necessary in this case to give flexibility so that the conductor can be handled and bent. Additionally, a number of the inner strands of transmission lines are made of steel. The strength of the steel is used to support the wires the long distances between transmission towers and to conduct electrical current. Aluminum is used on external strands for their lightweight characteristics. Standard cables are made with 7, 19, or 37 strands of wire.

Superconductors

Development of new types of materials allows electric current to flow at very low resistance. These materials are called *superconductors*, **Figure 2-6.** It has been learned that at extremely low temperatures, those near absolute zero (0 degrees Kelvin (K), –273 degrees Celsius (C), –460 degrees Fahrenheit (F)), the resistance of many materials drop to zero, or almost zero, ohms. These materials are called superconductors because they offer no resistance to electron movement. In this way, with reduced resistance, there is no loss in energy and electricity can be transferred much more efficiently. The goal, of course, is to find materials that do not need to be cooled to such low temperatures. Materials that superconduct at room temperatures or above will provide a revolution in such things as electric motor design and the speed at which computers operate.

Uses for Resistance

One might think that resistance is a very undesirable thing to have in an electrical circuit. This is not necessarily so.

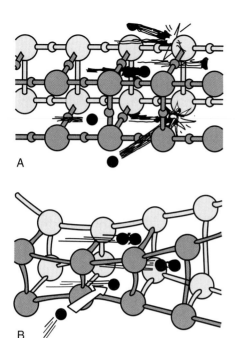

Figure 2-6.
These drawings show the difference in electron flow between a traditional conductor and a superconductor. A—In a traditional conductor, electrons can crash into the atomic structure and be slowed down. B—In superconductors, electrons fly right through materials, exhibiting no resistance, and, therefore, no loss in energy.

Resistance is often purposely introduced into circuits to produce desired results. Also, resistance is used to produce heat. The energy used up in a resistance device appears as heat. An electric range has resistance elements, which are the cooking surfaces on top of the stove. Incandescent lightbulbs contain resistance wire that get white hot and produce light.

The electrical circuits that you will construct in *Project 1—Experimenter* will help you to understand the flow of current through a resistance. In this project, you will use electricity to produce light. The switches demonstrate common methods of turning lights *on* or *off*.

Resistance in any circuit is the only component that uses up power and causes losses in the power source or supply. Consequently, any load on an electrical circuit that uses power can be represented by the symbol of a resistor.

The letter symbol for resistance is *R*. Its symbol in an electrical circuit is shown in **Figure 2-7**. Whenever you see the Greek letter *omega* (Ω) it will mean ohms.

Resistors can also be variable in value. Variable resistors are called **potentiometers** or **rheostats** depending on how they are used. See **Figure 2-8**.

Resistor Color Code

In your electronics studies, you will use many types of resistors. Many carbon type resistors have bright colored bands. These bands will tell you the value of the resistors in ohms. The resistor *color code*, **Figure 2-9,** is for your information.

To identify a resistor from the color code, hold the resistor in your hand so that the color bands closest to the end of the resistor point to the left. The first band color is the first number of the value. See **Figure 2-10**. The second band color is the second number of the value. The third band color tells you what factor to multiply the first two numbers by, in other words, the number of zeros to add to the value.

Figure 2-7.
Electrical symbol for resistance. This resistor has an opposition (resistance to electrical current) of 100 ohms.

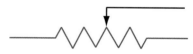

Figure 2-8.
Variable resistor symbol.

Color	Numerical Figure	Multiplier		Tolerance
Black	0	× 1	— add no zeros	
Brown	1	× 10	— add one zero	
Red	2	× 100	— add two zeros	
Orange	3	× 1000	— add three zeros	
Yellow	4	× 10,000	— add four zeros	
Green	5	× 100,000	— add five zeros	
Blue	6	× 1 Million	— add six zeros	
Violet	7	× 10 Million	— add seven zeros	
Gray	8	× 100 Million	— add eight zeros	
White	9	× 1000 Million	— add nine zeros	
Silver	—	× .01		± 10%
Gold	—	× .1		± 5%
None	—		———	± 20%

Figure 2-9.
Resistor color code.

Safety Suggestion!

When using appliance cord type of wire, remember that each little strand of the wire is a conductor. Frayed wires, or those not properly connected, can cause a dangerous situation. It is a proper safety practice to always coat the ends of a stranded wire with solder. This is called *tinning*. This keeps the fine wires in one group and allows you to make a good connection around a terminal.

For example, in Figure 2-10, a resistor with bands of brown, black, green, and silver represents 1,000,000 Ω or 1 megaohm (MΩ).

The fourth band, silver in this example, tells you how accurate the resistor must be in order to pass inspection. This is known as the resistor's **tolerance.** This resistor has a tolerance of ±10 percent. This means that the actual resistance of this resistor could measure as high as 1,100,000 ohms (1.1 megaohms) or as low as 900,000 ohms (900 kilohms) and still be acceptable (10 percent above to 10 percent below the specified value). The more accurate a resistor is, the more expensive it becomes. Most important equipment will use ±5 percent resistors and ±1 percent resistors.

Working with the color code is the best way to learn it. You will be given many chances to practice.

Quiz–Chapter 2

Write your answers to these questions on a separate sheet of paper. Do *not* write in this book.

1. The unit of quantity of electricity is _____.
2. The unit of electrical current is _____.
3. Potential difference is measured in _____.
4. Electromotive force is measured in _____.
5. Electrical pressure is measured in _____.
6. Electrical resistance is measured in _____.
7. Name four factors that affect the resistance of a conductor.
8. Resistance _____ as the wire length increases.
9. Resistance _____ as the wire's physical size decreases.
10. Draw the symbol for a resistance unit in a circuit.

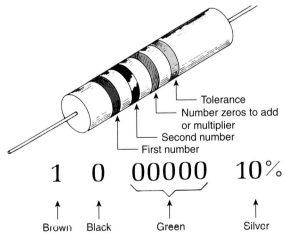

Figure 2-10.
How to read the resistor color code.

11. What is the resistance of 100 ft. of AWG No. 24 (0.5 mm) copper wire?
12. The resistance of a particular circuit wire cannot exceed one ohm. If the wire is 100 ft. long, what is the thinnest wire that can be used and still be less than one ohm?
13. What is meant by *tinning* stranded wires? Why should this be done?
14. Name two factors that must be considered by the power company when selecting a power transmission line.

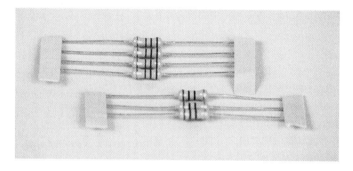

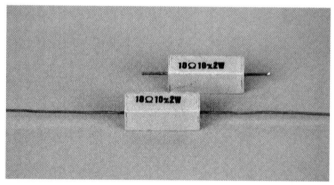

Selection of resistors. Top—Set of common 1/4 watt resistors. Bottom—Two higher wattage resistors. These resistors are safe to use up to 2 watts of power.

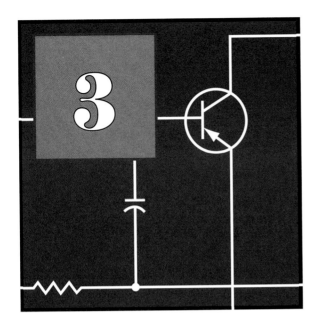

Meters, Reading a Meter

Objectives

After studying this chapter, you will be able to answer these questions:
1. What are the instruments used for measurement of values in an electric circuit?
2. What are the types and construction of meters?
3. How are meters connected in a circuit?

Important Words and Terms

The following words and terms are key concepts in this chapter. Look for them as you read this chapter.

ammeter	overload
autopolarity	reflecting galvanometer
autoranging	
iron-vane meter	series multiplier
milliammeter	shunt
ohmmeter	voltmeter

There are many types and kinds of meters used in electrical work. A meter can be considered as an indicating device for determining electrical quantities.

The three most widely used meters are:
1. *Voltmeter:* Used for measuring electrical pressure or force in volts (*always* connected in parallel).
2. *Ammeter:* Used for the measurement of current in a circuit in amperes (*always* connected in series).
3. *Ohmmeter:* Used for the measurement of resistance of a circuit in ohms (*always* used in a de-energized circuit).

Some meters can provide an *autoranging* feature. This means the meter will automatically adjust its range for small, medium, or large values. The meter in **Figure 3-1** is autoranging, as indicated by the very simple switching function. The large number of switch settings on the meter in **Figure 3-2** show that it is not autoranging. The meter in Figure 3-1 is digital (with number readouts instead of a moving pointer), which means that it is also autopolarity. *Autopolarity* means the meter will automatically display a negative (–) sign for a negative voltage or current. If values are positive, there will be no indication.

If too great a value of voltage or current is applied to a meter, it is said to be

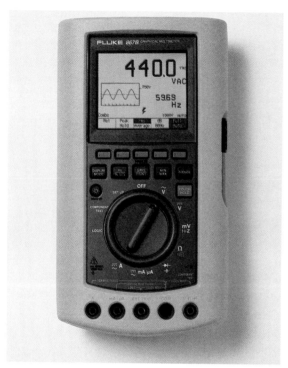

Figure 3-1.
This meter combines a digital readout with an analog wave readout. (Fluke Corp.)

Another special meter is the clamp-on meter, **Figure 3-3**. This device clamps around a wire carrying electrical current. By means of an inductive pickup (similar to those found on devices that detect the presence or absence of an object), the device can determine the amount of current flowing in a wire. This measuring device can be either digital or analog in nature. This meter also has the feature of measuring voltage and resistance with separate test leads.

The accuracy of measurement depends upon the skill of the operator and the quality of the instrument. Many industries and organizations have test instrumentation calibrated by electronic repair centers. These centers then compare their devices to standards supplied by the National Institute of Standards and Technology (NIST).

overloaded. An *overload* can damage or destroy a meter. Some meters indicate an overload by flashing their displays; others display the numeral 1. When an overload is indicated, the meter should be immediately disconnected from the circuit or the range switch quickly moved to a higher scale setting.

Other special purpose meters measure watts, watt-hours, frequency, capacitance, inductance, and other characteristics of circuits. Meters indicate the value of the measured electrical quantity on digital readouts or calibrated scales. As microelectronics (making circuits smaller and smaller) advances, many of these functions (frequency, capacitance, inductance along with transistor and diode checks) are being built into multimeters that are similar in size to the meters found in Figures 3-1 and 3-2.

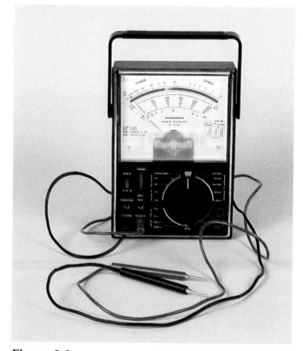

Figure 3-2.
This is an analog meter that measures amps, volts, and ohms. It is called a *VOM (volt-ohm-milliammeter)*.

Analog/Digital Meters

Two types of meters are in common use today. First, there is the analog or moving pointer meter. This type of meter has a moving pointer that moves through the lines on a calibrated scale or meter face. Secondly, there is the digital meter. The digital meter uses microelectronic circuitry to convert a current, voltage, or resistance into a signal that is then processed. A value of measurement is then displayed on the readout. Digital multimeters have decreased in price and now are less expensive than some analog meters. However, both types of meters are still in use and it will be important for you to learn how to use both. The good news is that the rules for connecting each type of meter are the same. However, at first, it is indeed easier to read the digital meter than the analog meter.

A time will come when analog meters will be a lot like the analog clock or wristwatch (second, minute, and hour hands) that has almost been totally replaced by digital clocks and watches. For the time being, knowing how to use both analog and digital meters is useful information.

> ## History Hit!
>
> Arsène (Jacques) d'Arsonval (1851–1940)
>
> d'Arsonval was born in Borie, France. In 1882, at the age of 31, he was made the director of the laboratory of biological physics at the College of France. He was the inventor of the **_reflecting galvanometer_** (a very sensitive meter used to measure current). The operating mechanism that uses a fixed magnet and moving coil is called the d'Arsonval meter movement in his honor.

d'Arsonval Meter

The basic movement used in electrical measuring devices consists of a fixed permanent magnet and a moving coil. Such a device was first used by the French physicist d'Arsonval in 1890.

The meter depends upon the interaction between a fixed magnetic field and a varying field. A coil of wire is suspended so it can turn within a fixed magnetic field. A current flowing through the coil in the proper direction creates a magnetic polarity that interacts with the fixed field and causes the moving coil to rotate. The amount of rotation is limited by tiny springs. These springs also return the movable coil to its original position when the coil is not magnetized. A pointer or indicating device can be fastened to the moving coil.

The rotation of the coil causes deflection (or movement) of the pointer along a dial or scale, which is calibrated in units of measurement. The force that causes the coil to

Figure 3-3.
A digital clamp-on meter. (Fluke Corp.)

move is proportional to the strength of the fixed magnetic field and the magnetic field of the coil. The fixed field remains constant, but the moving coil field is dependent upon the current flowing through the coil. The larger the current, the greater the rotation and deflection of the pointer. **Figure 3-4** is a phantom view (able to see inside) of a d'Arsonval meter movement.

Iron-Vane Meter

A second type of indicating device is the *iron-vane meter.* In this instrument, a fixed and a movable iron vane (a very thin piece of metal) are suspended within a coil. A pointer is attached to the moving vane. When a current flows in the coil, the two vanes become magnetized to the same polarity. As like poles repel each other, the moving vane rotates away from the fixed vane. The force causing the movement is proportional to the magnetic field of the current flowing through the coil. The deflection of the moving vane is limited by the force of a calibrated spring. Iron-vane meters can be made to measure current or voltage. A simplified sketch of this type instrument is shown in **Figure 3-5.** The iron-vane meter can withstand slight mechanical bumps without losing its accuracy.

Ammeters and Meter Shunts

If a meter is designed to measure amperes, it is called an ammeter. If it measures milliamperes, it is a *milliammeter.* A combination meter that reads either in milliamperes or amperes can be made by switching precision shunt resistors across the meter terminals. A *shunt* resistor is an alternative (parallel) path for current flow to take. In this case, the current bypasses, or is shunted around, the meter movement. If the resistance of the moving coil is known, and the current necessary for full scale deflection (maximum readable value) of the meter is also known, one can then calculate the necessary value of the shunt(s).

Figure 3-4.
Drawing showing the delicate inner workings of a d'Arsonval meter movement.

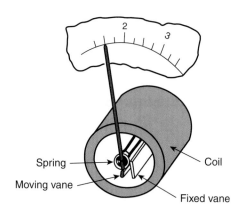

Figure 3-5.
A sketch of the rugged iron-vane meter movement.

Math Manipulation!

This next problem introduces Ohm's law, a formula that you will learn in depth in Chapter 4. A milliammeter has a scale that requires one milliampere (0.001 amp) for full scale deflection of the pointer. The internal resistance of the moving coil in the meter is 25 ohms. The voltage required for full scale reading of this meter would be:

$$E = I \times R$$

or

$$E = 0.001 \text{ A} \times 25 \text{ Ω} = 0.025 \text{ V}$$

If we wished to have the meter read from 0–10 mA, it would be necessary to connect in parallel with the meter a resistor shunt that would carry nine-tenths of the current (0.009 amps) leaving the remaining one-tenth or 0.001 amps for the meter to provide a full scale reading. **Figure 3-6** shows this circuit.

The shunt resistance can be found by:

$$R = \frac{E}{I} = \frac{0.025 \text{ V}}{0.009 \text{ A}} = 2.78 \text{ Ω}$$

When this shunt is used, the meter will read from 0–10 mA rather than 0–1 mA. A reading of 0.8 for instance, would mean 8 mA.

The range of the meter can be further increased to read 0–100 mA. In this case, 0.001 amp flows through the meter (for full scale deflection) and 0.099 amperes must flow through a shunt. The shunt resistance would be:

$$R_{shunt} = \frac{0.025 \text{ V}}{0.099 \text{ A}} = 0.253 \text{ Ω}$$

A circuit incorporating all three ranges is shown in **Figure 3-7**.

Switch Position	Scale
1	0–1 mA
2	0–10 mA
3	0–100 mA

Voltmeters and Multipliers

When the meter is used as a voltmeter, the range can be extended by *series multiplier* resistors.

Math Manipulation!

Using our previous example, we wish the meter to read 0–1 volt. As a current of 0.001 amps is required for full scale deflection, the total resistance in the circuit must be:

$$R_T = \frac{1 \text{ V}}{0.001 \text{ A}} = 1000 \text{ Ω}$$

As the resistance of the meter is 25 Ω, then the series resistor must equal 1000 Ω – 25 Ω = 975 Ω.

To increase the range to read 0–100 volts:

$$R_T = \frac{100 \text{ V}}{0.001 \text{ A}} = 100,000 \text{ Ω}$$

and series resistor equals:

$$100,000 \text{ Ω} - 25 \text{ Ω} = 99,975 \text{ Ω}$$

The diagram of this circuit with switching arrangement appears in **Figure 3-8**.

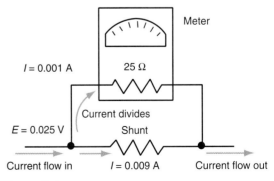

Figure 3-6.
The shunt resistor carries nine-tenths of the current, while the meter movement still obtains 1 mA. This is enough current for full-scale deflection of the pointer.

Safety Suggestion!

The circuit shown in Figure 3-7 must not be connected to the circuit when changing to higher range settings (0–10 mA *(Position 2)* to 0–100 mA *(Position 3)*). If the meter is connected to a current of greater than 1 mA when the switch is between settings (especially between positions 2 and 3), all current is directed to the meter. Between positions, the switch is making no contact, hence, a dead end for current. This would overload the meter, possibly bending the pointer as it goes over range. It can even burn out the coil, which would destroy the meter. Commercially available meters safeguard against this possibility.

Note: Ammeters measure current and are *always* connected in *series* with the circuit.

Safety Suggestion!

Voltmeters do not care if an open exists at the input lead to the meter. However, one and only one switch connection must be made at one time. That is to say, switch positions 1, 2, and 3 must never be made at the same time. This would reduce the overall resistance of the multiplier and supply too much current to the meter movement. Again, store purchased multimeters have this special feature already designed in.

Note: Voltmeters measure the difference of potential between two points and are *always* connected in *parallel* with the circuit.

In the multimeter, both the series multipliers and the shunts are connected to the basic meter by means of a switching arrangement, and the meter can be used to measure either voltage or current.

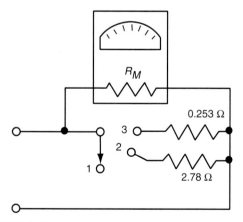

Figure 3-7.
A meter can have several ranges by adding shunt resistors to the circuit.

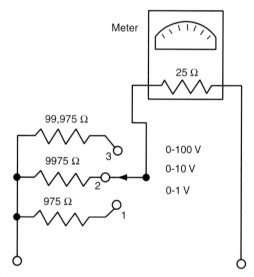

Figure 3-8.
The range of the voltmeter can be increased by adding multiplier resistors in the circuit.

Ohmmeters

In order to measure resistance with a meter, a variable resistor, (R_1), a voltage source, (B_1) and the unknown resistor to be measured, (R_X) are connected in series with the meter, as shown in **Figure 3-9.**

Math Manipulation!

Continuing with our same meter, the coil requires 0.001 amps for full scale deflection. By using two flashlight batteries in series as a voltage source (1.5 volts each), the total resistance of the circuit would be:

$$R_T = \frac{3\text{ V}}{0.001\text{ A}} \text{ or } 3000\text{ }\Omega$$

In other words:

$$R_M + R_1 = 3000\text{ }\Omega$$

Safety Suggestion!

As you have learned, ohmmeters supply their own energy, typically in the form of a battery. For this reason, ohmmeters must *never* be connected to an energized circuit. Any type of voltage/current combination added to the ohmmeter circuit will surely damage it. Be sure to "power down." *Never* use ohmmeters in an energized circuit.

R_1 is usually a variable resistor to compensate for the gradual changes in battery voltage due to usage and the amount of test lead resistance.

In order to use the meter, the terminals X and Z are shorted together and R_1 is adjusted to read zero resistance at full scale deflection of the meter. Then an unknown resistance can be inserted between X and Z terminals and its resistance can be read on a special calibrated scale of the meter. When points X and Z are open, the pointer does not deflect (move) and indicates infinite resistance on the scale.

Note that the ohmmeter scale is reversed. That is, the full scale deflection value is 0 ohms at the right side of the meter face. At the far left side of the meter face, the scale reads infinite. This is opposite to both voltage and current scales and is due to the nature of the circuit. There are other types of ohmmeter circuits you may wish to review, such as the shunt type of ohmmeter. However, this is a very common circuit for the moving coil, or d'Arsonval, meter movement.

Meter Connections

Remember, when measuring the current in a circuit, the ammeter must be inserted in series with the circuit. In other words, the circuit must be broken in order to connect the meter, **Figure 3-10**.

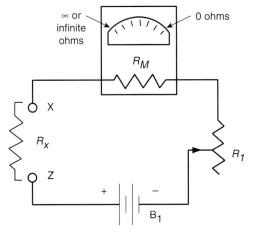

Figure 3-9.
Schematic diagram of a series ohmmeter.

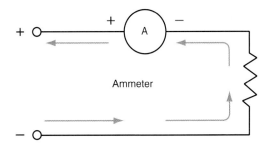

Figure 3-10.
The ammeter is connected in series to measure current.

Learning to Read a Meter

The Index Scale shown at the top of the drawing on page 31 is not a part of the regular meter. It is included on the drawing simply as a means of presenting meter-reading problems.

Practice in meter reading may be obtained by making use of the teaching aid shown on page 31.

On the Index Scale, each mark between the numbers equals two tenths. In solving the problems which follow, a ruler or a black thread is placed across the page from the dot in the center of the small circle at the bottom of page 31, to the index number at the top of the meter. The selector switch is assumed to be in the position specified in the problem. (Write answers to problems on a separate sheet of paper.)

I.	Selector Switch	300 volts dc		VI.	Selector Switch	× 10 ohms
	Index	Meter Reading			Index	Meter Reading
	3.0	--			4.0	--
	6.2	--			9.2	--
	10.0	--		VII.	Selector Switch	× 1000 ohms
	5.8	--			Index	Meter Reading
II.	Selector Switch	60 volts dc			2.0	--
	Index	Meter Reading			6.6	--
	2.0	--		VIII.	Selector Switch	× 100,000 ohms
	9.4	--			Index	Meter Reading
	7.4	--			8.0	--
	11.4	--			4.0	--
III.	Selector Switch	12 volts dc		IX.	Selector Switch	12 dc mA
	Index	Meter Reading			Index	Meter Reading
	4.4	--			9.0	--
	9.2	--			3.2	--
	5.8	--		X.	Selector Switch	1.2 dc mA
IV.	Selector Switch	3 volts dc			Index	Meter Reading
	Index	Meter Reading			9.0	--
	2.0	--			5.2	--
	11.0	--		XI.	Selector Switch	1.2 dc volts
	4.6	--			Index	Meter Reading
V.	Selector Switch	× 1 ohms			3.0	--
	Index	Meter Reading			8.0	--
	10.6	--			5.4	--
	4.8	--				

Chapter 3 Meters, Reading a Meter

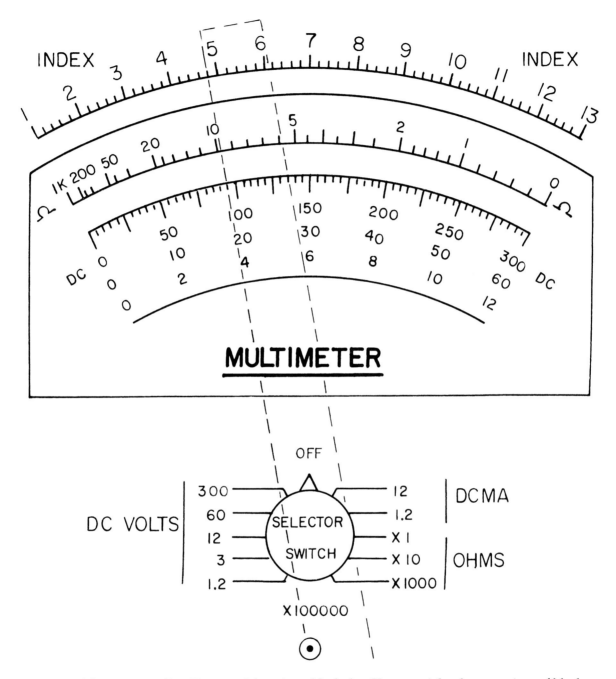

Teaching Aid for Meter Reading. To use, stick a pin at black dot. Use a straightedge or a piece of black thread from the dot to the index scale.

When measuring voltage, the meter is connected in parallel or across the circuit, **Figure 3-11.**

When using the ohmmeter, be sure there is no power applied to the component or the meter circuit, **Figure 3-12.**

Analog/Digital Meters— Which One Is Best?

As stated earlier, the digital multimeter is easier to use. Polarity does not matter with the digital meter (a negative sign will be displayed if the polarity is reversed). Many digital meters are autoranging, so you do not need to worry about setting the meter to the highest range when measuring an unknown voltage or current. In fact, there are even some digital meters that talk to you—they indicate the value on a display or readout and tell you the value verbally, with a computer generated voice! In the past, analog meters were much less costly than digital meters. However, their cost has come down, as even their functions have increased.

It seems wise, for a beginner in the field of electricity/electronics, that he or she learns on a durable, relatively inexpensive piece of equipment. Remember, one wrong reading may destroy a multimeter forever— the internal fuse may not act in time! For this reason, the analog meter may be a good choice for a first time user. The digital meter is easier to use, it is even more accurate, but it may not be as easily replaced if ruined.

Meter Precautions

1. A meter is a delicate precision instrument. Do *not* drop it or treat it roughly.
2. Be sure all switches are correctly set for range and type of measurement.
3. When making unknown measurements, *always start with the highest range on the meter.* If you tried to measure 100 volts on the one volt range, the meter could be destroyed. This holds true for both voltage and current readings! For resistance readings, it does not matter where you start the range settings, just be sure the circuit is de-energized.
4. When connecting a meter, be sure to observe the proper polarity. *Negative (black lead) to negative, positive (red lead) to positive.*
5. When measuring current *always* connect the ammeter in series with the circuit.
6. When measuring voltage *always* connect the voltmeter parallel to the circuit.

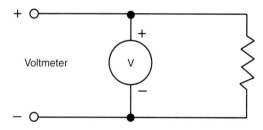

Figure 3-11.
The voltmeter is connected in parallel to measure voltage.

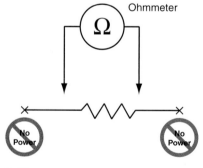

Figure 3-12.
Be sure the ohmmeter is never connected to an energized circuit.

7. Do *not* attempt to measure resistance in a circuit if the circuit is "alive" (the power is turned on) Figure 3-12. This will certainly damage the ohmmeter! Remove all power from the circuit to measure resistance.

Quiz–Chapter 3

Write your answers to these questions on a separate sheet of paper. Do *not* write in this book.

1. Draw the diagram for an ammeter with three ranges of current measurement.
2. If this meter required 0.001 amp for full scale deflection and the resistance of the meter coil was 100 ohms, compute shunt resistors for:
   0–10 mA range _____ Ω
   0–100 mA range _____ Ω
3. Draw diagram for a voltmeter with three ranges.
4. Compute series multiplier resistors: R_M = 100 Ω, 0.001 amp for full scale deflection.
   0–1 volt range _____
   0–10 volt range _____
   0–100 volt range _____
5. Draw the circuit for a simple series ohmmeter.
6. The two types of meter movements we have studied are the _____ and the _____ type meter.
7. Ammeters are connected in _____ with the circuit and voltmeters are connected in _____ with a circuit.

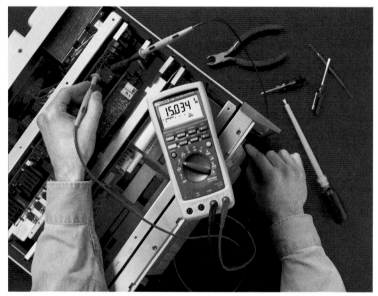
A multimeter is an essential tool for every electronics/electrical technician. (Reproduced with Permission—Fluke Corp.)

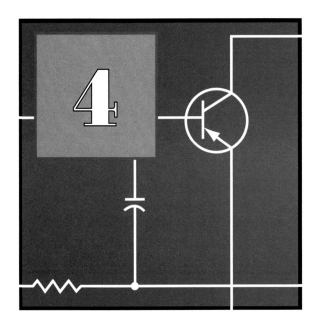

Ohm's Law

Objectives

After studying this chapter, you will be able to answer these questions:
1. What is the relationship between voltage, current, and resistance in a circuit?
2. What is Ohm's law and how can we use it to solve electrical circuit problems?
3. What types of switches are used in electrical circuits?

Important Words and Terms

The following words and terms are key concepts in this chapter. Look for them as you read this chapter.

conductive pathway
control
electrical circuit
load
Ohm's law
voltage source

The Simple Circuit

Electrical circuits are complete pathways through which electric current flows. Three elements are basic to all circuits:

1. *Voltage source* (such as a battery or generator). A device that supplies the energy.
2. *Load* (such as a resistor, motor, or lamp). A device that uses energy from the voltage source.
3. *Conductive pathway* (such as an insulated wire or printed circuit board). A path from voltage source to load and back, which carries electrical current.

Circuits usually contain a fourth element, as well. A *control* device such as a switch, fuse/circuit breaker, or relay may be used to stop, start, and/or regulate the flow of electricity.

Figure 4-1 is a schematic diagram of a simple circuit. The symbols show that the circuit has a battery for the voltage source; a load device (resistor), which uses the energy from the voltage source; an "on/off" switch; and connecting wire to conduct the current.

Ohm's Law

One of the basic laws of electrical circuits is Ohm's law. This shows mathematically the relationship between voltage (E),

36 Electricity

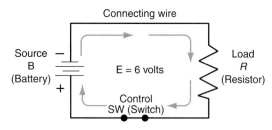

Figure 4-1.
A simple electrical circuit. Note connection of elements that make up circuit.

---

current (*I*), and resistance (*R*). A thorough understanding of the use of Ohm's law will help you to understand how any circuit operates.

If you do not completely understand Chapter 2, perhaps now is a good time to review it once again. You will remember that an electric current was caused to flow in a conductor when a force or voltage was applied to the circuit. Figure 4-1 shows a simple circuit using a battery as a voltage or potential difference source.

R represents the resistance in the circuit and *I* stands for "intensity" of the current. *E* or *V* represents electromotive force.

As the voltage of battery (B) is fixed and the resistance of the circuit is fixed, a definite value of current will flow in the circuit. (Note the direction of current flow as indicated by the arrows.)

If the voltage were increased to twice the value, as in **Figure 4-2,** then the current would also increase to twice its former value. As the voltage increases, the current

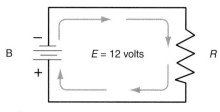

Figure 4-2.
As voltage is increased, the current increases.

---

increases. As the voltage decreases, the current decreases. A mathematician would say that the current and voltage are in *direct proportion* to each other.

The current flowing in these circuits also depends on the resistance of the circuit. If we increase the resistance to twice its value, the current is cut in half. We may conclude that as the resistance increases, the current decreases. As the resistance decreases, the current increases. Again, mathematically speaking, the current is in *inverse proportion* to the resistance.

Georg Simon Ohm, the German scientist, proved this relationship to be true in his experiments. The law is named in his honor. **Ohm's law** is stated as:

$$I = \frac{E}{R}$$

where,
I = current in amperes
E = voltage in volts
R = resistance in ohms

By simple algebra, the formula may be changed to read:

$$R = \frac{E}{I} \text{ or } E = IR$$

Math Manipulation!

One may readily see that if any two quantities are known in a circuit, the third quantity can be found. Referring to **Figure 4-3**, notice that values have been assigned to *E* and *R*.

The current is easily computed by Ohm's law:

$$I = \frac{E}{R} \text{ or } I = \frac{6 \text{ V}}{12 \text{ }\Omega} = 0.5 \text{ A}$$

If the voltage were unknown and we knew the current and resistance: $I = 0.5$ A, and $R = 12$ ohms, then:

$$E = I \times R \text{ or } 0.5 \text{ A} \times 12 \text{ }\Omega = 6 \text{ V}$$

If the resistance were unknown and the voltage and current were given as: $I = 0.5$ A, $E = 6$ volts, then:

$$R = \frac{E}{I} \text{ or } \frac{6 \text{ V}}{0.5 \text{ A}} = 12 \text{ }\Omega$$

If you have difficulty remembering this equation in its three forms, the simple memory device shown in **Figure 4-4** may help.

Place your finger over the unknown quantity and observe what it equals. For example: Put your finger over E, the answer is $I \times R$. Put your finger over I, the answer is:

$$\frac{E}{R}$$

Put your finger over R, the answer is:

$$\frac{E}{I}$$

The purpose of the memory device is to make it easier to remember how to use Ohm's law. The best way to learn Ohm's law is to practice its use.

You must remember that when using Ohm's law, E, I, and R must be in volts, amperes, and ohms, respectively. Study

History Hit!

Georg Simon Ohm (1787–1854)

Ohm was educated at the University of Erlangen and became a professor of physics at Munich in 1849. Ohm developed the law for which he is best known. However, it should be noted that he received little acknowledgment for this achievement for 20 years. He also did much pioneering research on the human ear and how different sounds or frequencies are broken down by the different parts of the inner ear.

Figures 4-5 and **4-6**. Frequently current is given in milliamperes, which is:

$$\frac{1}{1000} \text{ of an ampere or } 0.001 \text{ A}$$

You must convert to amperes before using the equation. Studying the following examples will help you to do this:

1 ampere = 1000 milliamperes
0.5 ampere = 500 mA
0.1 amp = 100 mA
50 mA = 0.05 amp
500 mA = 0.5 amp
10 mA = 0.01 amp
1 mA = 0.001 amp

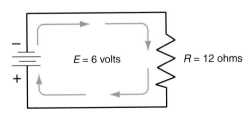

Figure 4-3.
The current equals 0.5 amperes.

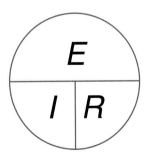

Figure 4-4.
A memory device for Ohm's law.

1 milliampere (mA) = 1/1,000 (.001) ampere
1 microampere (mA) = 1/1,000,000 (.000001) ampere

Figure 4-5.
Prefixes used in measuring current.

Math Manipulation!

As you review Figures 4-5 and 4-6, realize that the powers of 10 make it easier to work with very large or very small numbers. These units (milliamperes and microamperes) make it much easier to list very, very small units of current. In the future, we will work with much larger units. It will be important for you to use the powers of 10 in electricity and electronics, so closely review Figures 4-5 and 4-6 as you examine these mathematics principles.

Overload Protection of Circuits

It should be quite clear that a certain kind and size of wire has a specified ability to conduct an electric current. All conductors have some resistance. When a current overcomes this resistance, heat is produced. If a wire is operated within its limitations, this heat is dissipated in the surrounding air and its temperature does not rise excessively. However, if too great a current is forced through the conductor, the temperature will rise to a point where the wire will become hot. If the wire gets hot enough, the insulation that surrounds the conductor may melt off. The wire may even get hot enough to melt itself and be destroyed. If it is near combustible material, such as in the wall of your home, a fire might result.

Overloading a circuit can occur from two causes:
1. An excessive load that draws beyond a safe amount of current.
2. A direct (sometimes called a dead) short circuit.

Circuits and appliances are usually protected by a fuse or circuit breaker. A *fuse* is simply a thin strip of metal that melts at a low temperature. Those used in the home are usually designated 15 and 20 amperes. (Carefully examine your fuse box or circuit breaker panel at home.) Note the Safety Suggestions later in this chapter. Typically, circuit breakers have replaced fuses in most home and industrial applications. However, older installations may still contain fuses, or a combination of fuses and circuit breakers.

If a current exceeds the fuse rating, it will melt and open the circuit, preventing damage of equipment and danger of fire. The symbol for a fuse in electrical circuit diagrams is shown in **Figure 4-7**.

Some fuses, of course, are made to carry heavier currents. You will generally find a main power panel in a home rated between 100 and 200 amperes. In fact, as homes utilize more and more electricity, even larger amounts of current are

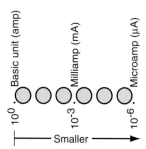

Figure 4-6.
Conversion chart for ampere prefixes.

Figure 4-7.
The symbol for a fuse.

> **Safety Suggestion!**
>
> If you do look into the power panel of your home, be careful! Be sure to touch only the panel door to open it and close it. Fuses or circuit breakers may have been removed. This can expose dangerous and fatal electrical connections! If there are loose wires or metal conductors exposed, do not touch the panel. If you are unsure of the safety of the panel box, leave it alone. You may want to have a qualified electrician examine this possibly dangerous situation.

> **Safety Suggestion!**
>
> If you were to examine the cords to lamps and appliances at home, it is possible that you may find a damaged one. Needless to say, these electrical cords are dangerous. You can receive a serious burn from a short circuited lamp cord, as well as the danger of an electric shock. It is best to refer these problems to a qualified repair person and/or electrician.

provided at the main power panel. You will often find in the electrical service panel in your home fuses of the cartridge type, rated for 100 to 200 amperes or more. These main fuses carry the total current used by all of the circuits in your home and give further protection.

In **Figure 4-8,** a simple load in the form of a resistor is connected across a voltage source. If the insulation should become worn or frayed so that wire A could touch wire B, the sparks would fly. This is called a *short circuit.* This can happen with lamp and appliance cords.

One improved safety device is called the circuit breaker. You will study these in detail in the unit on magnetism. A *circuit breaker* is a magnetic or thermal device that automatically opens the circuit when an excessive current flows. See **Figure 4-9.** Circuit breakers must be manually reset before the circuit can be used again. Circuit breakers have three positions, *on, off* and *tripped.* The tripped position has stopped current because of an overcurrent condition (a short circuit or a current above its rated value). Needless to say, the cause of the overload should be investigated and removed or repaired before the current is turned on again. Circuit breakers are replacing fuses in most applications in the home and industry. One reason is because they are reusable, the fuse is not. As you examine your home or school electrical service panels, are fuses or circuit breakers used?

Circuits and Switches

There are many varieties of switches used in electrical equipment. The student should be familiar with the common types, **Figure 4-10.**

If only one wire or one side of a line is to be switched, the single-pole, single-throw

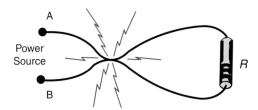

Figure 4-8.
A short circuit, sometimes called a direct or dead short, across the power supply connections A and B will make the sparks fly!

Figure 4-9.
Symbol for a circuit breaker.

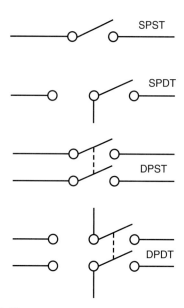

Figure 4-10.
Diagrams of switch types.

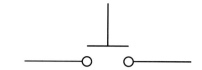

Figure 4-11.
Push button normally open (PBNO).

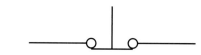

Figure 4-12.
Push button normally closed (PBNC).

(SPST) switch is used. If both sides of the line were to be switched, then a double-pole, single-throw (DPST) switch would be used. Reference to Figure 4-10 will show the circuit diagrams for various switches. If a single line is to switch first to one point and then to another, the SPDT switch can be used. If a double line is to be switched to two other points, then the DPDT switch would be used.

Another type of switch used is the push-button switch. These switches can be of two main types: normally open or normally closed. The normally opened push button (PBNO) is found in **Figure 4-11.** This switch only completes a circuit when pushed. **Figure 4-12** is the opposite of PBNO, the normally closed push button (PBNC). This switch opens the circuit only when pushed or depressed. Otherwise, the circuit is closed. The term "normally" is used to indicate that the push-button switch is at rest (untouched by someone).

Frequently, it is desirable to switch a circuit from two different locations. In this case, a three-way switch is used. Perhaps you have such switches in your home that permit you to turn a light on or off from two places in a room or hallway. The schematic diagram of this circuit is shown in **Figure 4-13.**

The light is on, but it can be turned off by moving either switch 1 or 2. In **Figure 4-14** the light is off, but it can be turned on by either switch 1 or 2. Follow the circuit through the switches in each position.

In *Project 1—Experimenter, Problem 6,* you will gain first-hand experience in connecting three-way switches, so that lights can be controlled from different locations.

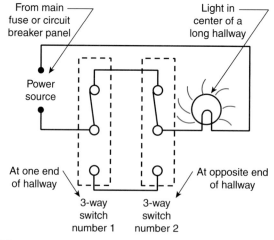

Figure 4-13.
A three-way switch circuit. The light is on.

Chapter 4 Ohm's Law

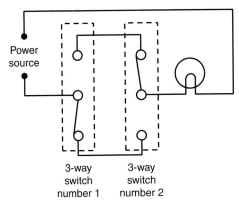

Figure 4-14.
A three-way switch circuit. The light is off.

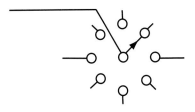

Figure 4-15.
Multiple-pole rotary switch.

A common type of switch used in electrical equipment is the multiple-pole rotary switch. In this switch, the rotary contact arm can be connected to one side of an electrical circuit and the contacts can connect to several other circuits. By turning the control knob, any desired circuit configuration can be used. Multiple switching operations can be done by mounting several rotary switches on one control shaft, **Figure 4-15.** Where might a switch like this be used?

Quiz–Chapter 4

Write your answers to these questions on a separate sheet of paper. Do *not* write in this book.
Draw the circuit diagram for each of the following problems and compute the unknown quantities.

1. A circuit has an applied voltage of 100 volts and a resistance of 1000 ohms. What is the current flowing in the circuit?
2. A circuit that contains 100 ohms resistance has a current of two amperes. What is the applied voltage?
3. A circuit that contains 10,000 ohms of resistance has a current of 100 mA. What is the applied voltage?
4. A circuit has an applied voltage of 200 volts that causes a 50 mA current to flow. What is the circuit resistance?
5. An applied voltage of 50 volts causes a current of 2 amperes to flow. What is the circuit resistance?
6. A voltage of 500 volts is applied to a circuit that contains 100 ohms of resistance. What is the current?
7. If applied voltage is 400 volts and resistance is 20,000 ohms, what is the value of I?
8. A meter indicates a current flow in a circuit of 0.5 amp. The circuit resistance is 500 ohms. What is the value of E?
9. What applied voltage will cause 500 mA of current to flow through 500 ohms of resistance?
10. What applied voltage will cause 10 mA of current to flow through 1000 ohms of resistance?
11. An electric appliance has a resistance of 22 ohms. How much current will it draw when connected to a 110 volt line?
12. A 110 volt house circuit is limited to 15 amperes by the fuse in the circuit. The following appliances are connected to the circuit. Compute the individual currents for each appliance. What is the total current flowing in the circuit? Will the fuse permit this current to flow?
    Appliance 1 draws 2 amperes.
    Appliance 2 has a resistance of 40 ohms.
    Appliance 3 has a resistance of 20 ohms.

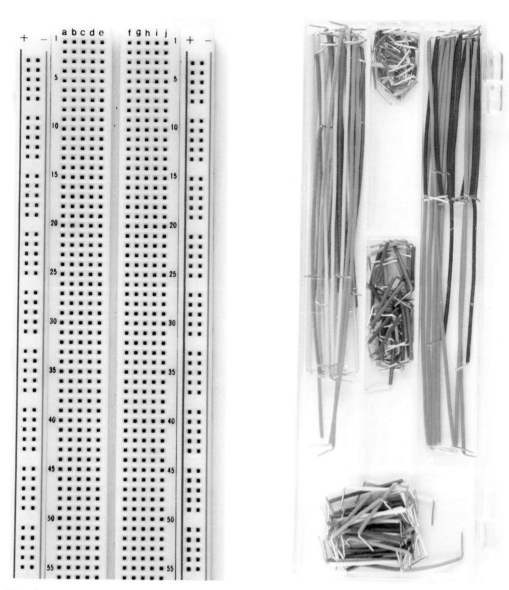

Hands-on experiments are an important part of learning electricity and electronics. Thousands of different circuits can be created on generic circuit boards. Bus wire and jumper wire are often used to make connections.

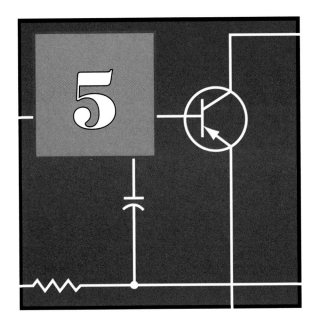

5

Power

Objectives

After studying this chapter, you will be able to answer these questions:
1. What is power?
2. What is the relationship between Ohm's law and the power law?

Important Words and Terms

The following words and terms are key concepts in this chapter. Look for them as you read this chapter.

energy	power law
horsepower	watt (W)
kilowatt-hour	Watt's law
PIRE wheel	watt-hour
power (P)	work

Power

The rate at which electrical energy is delivered to a load is called electric *power (P)*. The unit of measurement of electric power is the *watt (W)*. In electrical circuits, power consumed is equal to the current multiplied by the voltage or:

$$P = I \times E$$

Therefore, one watt of power is the result of one ampere of current driven by a one volt force through a circuit.

If a circuit with one volt of electrical pressure causes one ampere of current to flow for one hour, then one *watt-hour* of electrical energy has been used. A watt-hour is a relatively small unit of energy. You should be more familiar with the *kilowatt-hour*, which means that energy is used at the rate of 1000 watts per hour.

To convert between watts and kilowatts, it may be helpful to review the following examples.

$$
\begin{aligned}
1000 \text{ watts} &= 1 \text{ kilowatt (1 kW)} \\
500 \text{ watts} &= 0.5 \text{ kW} \\
100 \text{ watts} &= 0.1 \text{ kW} \\
50 \text{ watts} &= 0.05 \text{ kW} \\
5 \text{ watts} &= 0.005 \text{ kW} \\
1 \text{ watt} &= 0.001 \text{ kW}
\end{aligned}
$$

44 Electricity

```
1000 watts      = 1 kilowatt (kW)
500,000 watts   = 500 kW or .5 megawatt (MW)
1,000 kW        = 1,000,000 watts or 1 MW
```

Figure 5-1.
Prefixes used in calculating power.

It may be helpful to review **Figure 5-1**, showing prefixes used when calculations are made in power, and **Figure 5-2**, a conversion chart for wattage prefixes. When you pay your electric bill, you will notice that you are paying for energy used at so much per kWh, or kilowatt-hour. A lightbulb in your room may be rated at 100 watts. To keep your light burning for one hour would require 100 watt-hours of electrical energy, and in ten hours it would use one kilowatt-hour of electrical energy.

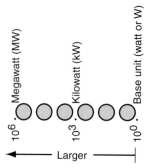

Figure 5-2.
Conversion chart for power prefixes.

Math Manipulation!

A toaster uses eight amps of electricity at 110 volts pressure. How much power does it consume?

Power equals volts times amperes, so:

$$P = 110 \text{ V} \times 8 \text{ A} = 880 \text{ W}$$

If you should toast bread for a whole hour, the energy used would equal 880 watt-hours or 0.88 kilowatt-hours.

Figure out how much it costs for the electricity to iron shirts for one hour with an electric iron using 5 amperes of current.

The power consumed would be:

$$P = 110 \text{ V} \times 5 \text{ A} = 550 \text{ W}$$

So, in one hour the iron would use 550 watt-hours, or 0.55 kWh. If electricity costs 9 cents per kWh, the ironing would cost:

$$0.55 \text{ kWh} \times 9 \text{ cents} = 4.95 \text{ cents per hour}$$
(almost 5 cents per hour)

As power is the product of the voltage and the current in a circuit, one can measure these values by meters and compute the power of the circuits. It is easier to use a special meter that reads directly in kilowatt-hours as no computation is necessary. Such a meter is on the outside of your house, at the electrical service entrance.

Look at your home electric meter found near where the wires from the utility pole attach to your house. Should you have an underground service entrance (no utility poles), you will need to look around your house to locate it. Notice that some of the meters have a moving wheel that displays the rate of energy usage. Faster wheel movement indicates that more energy is in use. When does the wheel move fast? When does the wheel move more slowly? Also, more and more digital kilowatt-hour meters are in use. These have replaced many of the older mechanical movement type of kilowatt-hour meters.

By simple algebra we can write the *power law* or *Watt's law* in three ways as we did for Ohm's law:

$$P = I \times E$$

$$I = \frac{P}{E}$$

$$E = \frac{P}{I}$$

Chapter 5   Power

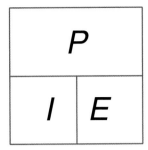

Figure 5-3.
A memory device for Watt's law.

The memory device shown in **Figure 5-3** will help you. Place your finger over the unknown quantity and observe what to do with the known two values in the formula.

With only a bit more algebra, Watt's law and Ohm's law can be combined to create equations that solve for unknown voltage, current, resistance, or power. **Figure 5-4** shows the complete set of equations resulting from combining these laws.

Figure 5-5 shows a memory device known as the **PIRE wheel** (Power, Current, Resistance, Voltage). It will help you solve problems that involve watts (P), amperes (I), ohms (R), and volts (E). If the values for

$I = \dfrac{E}{R}$ $R = \dfrac{E}{I}$

$I = \dfrac{P}{E}$ $R = \dfrac{E^2}{P}$

$I = \sqrt{\dfrac{P}{R}}$ $R = \dfrac{P}{I^2}$

$E = \dfrac{P}{I}$ $P = I \times E$

$E = I \times R$ $P = I^2 R$

$E = \sqrt{PR}$ $P = \dfrac{E^2}{R}$

Figure 5-4.
Equations used to figure unknown voltage, current, resistance, and power.

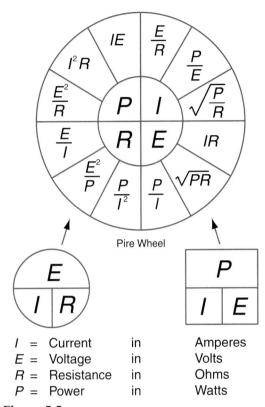

I = Current   in   Amperes
E = Voltage   in   Volts
R = Resistance   in   Ohms
P = Power   in   Watts

Figure 5-5.
A device that shows the formula resulting from the combination of Ohm's law and Watt's law.

any two terms are known, the other two values can be easily found.

Horsepower

We can define *energy* as the capacity for doing work. You are asked to mow the lawn. You are much more comfortable sitting in a lawn chair with a cool drink and a good book. However, you have the energy stored within you to cut the grass. You have the force to push the mower. When you push a lawn mower (apply a force over a certain distance), work is done. **Work** is the product of F (force) $\times D$ (distance) and is measured in foot-pounds. If the lawn mower requires a ten pound force to move it and you push it one hundred feet, then you have done 10×100 or 1000 foot-pounds of work.

But how long did it take you to do this work? What is the rate at which you work? This would be your power. Your power can be computed by the formula:

$$P = \frac{\text{work}}{\text{time}}$$

Math Manipulation!

If you took one minute to push the mower 100 feet, your power should be:

$$P = \frac{1000\ (F \times D)}{1\ \text{minute}} = 1000\ \text{foot-pounds per minute}$$

An employer may not be satisfied with the rate that you are doing the work. You are shown how to do the job by pushing the mower 100 feet in only 30 seconds or one-half minute. This power would be:

$$P = \frac{1000\ (F \times D)}{0.5} = 2000\ \text{foot-pounds per minute}$$

The same amount of work has been done, but due to greater power, the work is completed in less time.

You have heard of the term *horsepower*. When James Watt, the inventor of the steam engine, was looking for a suitable way to measure power, the horse was the common source of power. He compared the power of his engine to an ordinary horse. He might have used an oxen or a dog. If he had, we might be rating our automobiles today in oxen power or dog power. James Watt found that an average workhorse might work at the rate of 550 foot-pounds per second or 33,000 foot-pounds per minute. This rate of doing work is considered as one horsepower. Scientists have determined that one horsepower is equivalent to 746 watts of electrical power.

History Hit!

James Watt (1736–1819)

Watt was the son of a shipbuilder in England. Due to poor health, he had little formal education, however he became an instrument maker. While repairing a certain type of steam engine, Watt realized that he could greatly improve its overall efficiency. It was his improved steam engine that became so important to the Industrial Revolution. For this reason, James Watt was honored with his name being attached to the measurement of the rate of energy power production.

Earth and the Environment: How to Conserve Electrical Power

It is important to use all of our resources wisely, including electricity. As you will learn in further chapters, electricity can be generated or created in a variety of ways. However, all the techniques utilized widely today to generate electrical power are limited. The following list contains some tips on how to conserve electrical power in the home. These suggestions cost very little and only require a small bit of effort on your part. If used, they can save quite a bit of money. Even more importantly, they will help to save and preserve our Earth and its resources—for today and tomorrow.

- Adjust furnace thermostats downward and replace or clean filters regularly.
- Install a low-flow showerhead and similar devices made for faucets.
- Reduce hot water temperature and usage. (This also reduces the possibility of burns.)
- Install a water heater insulation blanket to reduce water-heating costs.

Chapter 5   Power

- When on vacation, reduce heating and air-conditioning if your residence is unoccupied.
- Seal air leakage to attics, basements, and other unconditioned spaces. Caulk around doors and windows.
- Repair leaks in water and steam pipes and in heating and cooling ducts.
- Discontinue operating second refrigerators and freezers if under utilized.
- Turn down waterbed heaters, insulate mattress and frame. An inexpensive time-clock can be used to prevent unnecessary daytime waterbed heater operation.
- Use lower wattage lightbulbs. Switch to fluorescent bulbs, which use less energy and last longer. Screw-in fluorescent fixtures make it easy to switch.
- Connect incandescent lights to dimmer switches. Use timers for security lighting.
- Close the fireplace damper when the fireplace is not in use.
- Wear warm clothing indoors in winter and light, loose fitting clothes in summer.
- Heat your house to no more than 68 degrees if health permits. Lower the temperature at night.
- In summer, set the air-conditioning at a minimum of 78 degrees. Install awnings or stick-on window screening materials. Shade the air conditioner's condensers, but do not restrict airflow. Plant deciduous trees and shrubs for shade.

Electric Motors

Electrical energy can be changed to mechanical energy. A motor is a good example of this conversion. Try to think of as many uses of the electric motor as you can. Electric motors are found in CD players (turning the disc), ceiling fans, heating/cooling equipment, computers (cooling fans and CD drives) and printers (moving the paper and the print head, if so equipped). There are many, many types of motors. Some are of general use (a simple ac or dc motor), while some are of special use (the stepper motor used in robotic applications).

All motors have a number of specified ratings or values that describe the motor. Common ratings include operating voltage, current (ac or dc), power (usually rated in horsepower), speed, and among others, torque. *Torque* is the force produced by a turning shaft (used in automobiles as well). If possible, examine some motors and review their nameplate data labels. These labels are found on motors and describe the characteristics of the motor. Motor horsepower can range widely. Motors such as the one used in a CD player have just a tiny fraction of a horsepower. Motors used in large industrial facilities can be rated in the hundreds of horsepower, and they can be as large as a good size office desk!

> ### Web Wanderings!
>
> http://www.si.edu/
>
> Check out the Smithsonian Web site for a rich resource of information related to science and technology. This site features a listing of events, and museum exhibitions. Subscriptions to "Smithsonian Focus," a free monthly electronic newsletter, are offered. Online events and multimedia on fascinating scientific and technological topics are also available.

Quiz–Chapter 5

Write your answers to these questions on a separate sheet of paper. Do *not* write in this book.

1. Power is equal to _____ × _____.
2. The rate at which energy is applied to a load of work is called _____.

3. Your house has a 110 volt electric circuit. How much current will a 1000 watt appliance use?
4. At 9 cents per kWh, how much will it cost to use the appliance in Question 3 for two hours?
5. One horsepower equals _____ watts.
6. A kilowatt equals _____ watts.
7. A certain appliance uses two amperes of current when connected to a 100 volt source. What is the power?
8. What is the resistance of the appliance in Question 7?
9. A certain appliance has a resistance of 100 ohms. At 100 volts how much current is used?
10. A meter used to measure energy consumed is called a(n) _____.

On a separate sheet of paper, copy the PIRE wheel in Figure 5-5. Work out the answers to the problems below using your PIRE wheel. Write your answers on a separate sheet of paper.

11. $E = 100$ V, $I = 2$ amps, $R = $ _____.
12. $E = 50$ V, $R = 1000$ ohms, $I = $ _____.
13. $I = 0.5$ amps, $R = 50$ ohms, $E = $ _____.
14. $E = 10$ V, $I = 0.001$ amps, $R = $ _____.
15. $I = 0.05$ amps, $R = 1000$ ohms, $E = $ _____.
16. $P = 10$ W, $I = 2$ amps, $E = $ _____.
17. $E = 100$ V, $I = 0.5$ amps, $P = $ _____.
18. $P = 500$ W, $E = 250$ V, $I = $ _____.
19. $I = 0.01$ amps, $R = 100$ ohms, $E = $ _____.
20. $P = 100$ W, $I = 2$ amps, $R = $ _____.
21. $E = 10$ V, $P = 10$ W, $R = $ _____.
22. $E = 500$ V, $I = 2$ amps, $R = $ _____.
23. $E = 100$ V, $R = 1000$ ohms, $P = $ _____.
24. $I = 0.5$ amps, $R = 50$ ohms, $P = $ _____.
25. $I = 4$ amps, $R = 10$ ohms, $P = $ _____.
26. $I = 10$ mA, $E = 50$ V, $P = $ _____.
27. $I = 20$ mA, $E = 100$ V, $R = $ _____.
28. $P = 10$ W, $I = 1$ amp, $R = $ _____.
29. $E = 1000$ V, $R = 1000$ ohms, $I = $ _____.
30. $I = 100$ mA, $R = 100$ ohms, $E = $ _____.
31. $I = 100$ mA, $R = 100$ ohms, $P = $ _____.
32. $P = 500$ W, $E = 100$ V, $I = $ _____.
33. $E = 100$ V, $R = 100$ ohms, $P = $ _____.
34. $E = 50$ V, $R = 10$ kilohms, $I = $ _____.
35. $P = 10$ W, $R = 10$ ohms, $E = $ _____.
36. $P = 50$ W, $R = 2$ ohms, $I = $ _____.
37. $R = 100$ ohms, $P = 100$ W, $E = $ _____.
38. $I = 0.001$ amps, $R = 1$ megaohm, $P = $ _____.
39. $E = 200$ V, $I = 200$ mA, $P = $ _____.
40. $E = 600$ V, $I = 300$ mA, $P = $ _____.

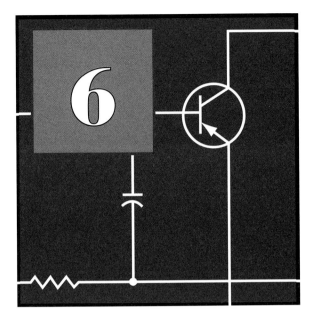

6

Series Circuits

Objectives

After studying this chapter, you will be able to answer these questions:
1. How are electrical devices connected in series?
2. What are the laws of series circuits?

Important Words and Terms

The following words and terms are key concepts in this chapter. Look for them as you read this chapter.

IR drop

Kirchhoff's current law

Kirchhoff's voltage law

series

voltage drop

Consider the example of a highway between two towns. This highway has four narrow bridges connected by stretches of road, as shown in **Figure 6-1.** All the traffic moves in one direction and must cross the bridges one after the other. The bridges can be considered to be in *series* with each other.

Before each bridge is a traffic sign, "Slow down, narrow bridge." You reduce your speed because the bridge has slowed down your progress.

You can compare this traffic situation to a simple electrical circuit, **Figure 6-2.** The highway has been replaced by an electrical conductor or wire. The bridges have been replaced with resistors, R. Your travel in the direction of the arrows has been replaced by a flow of electrons from negative to positive terminals of the power source. R_1, R_2, R_3, and R_4 are all in series with the power source. Each R has a certain resistance to the

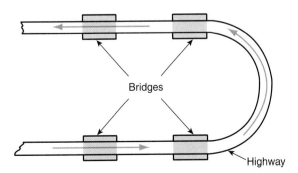

Figure 6-1.
The bridges on the highway are in series.

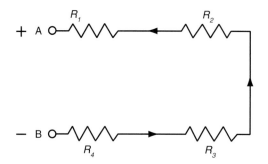

Figure 6-2.
The resistors are in series.

flow of electrons between points A and B. The total resistance in the *series circuit*, then, is equal to the sum of all the resistors.

$$R_T = R_1 + R_2 + R_3 + R_4$$

where R_T equals the total resistance.

As you study Figure 6-2, you observe that all electrons entering the circuit at point B, must flow through R_4, R_3, R_2, and R_1 before reaching point A. Also, all electrons going in at point B must come out at point A. None are lost along the way. So one can conclude that the electron flow, or current, must be the same at any point in the circuit.

$$I_T = I_{R_1} = I_{R_2} = I_{R_3} = I_{R_4}$$

where I_T is the total current of the circuit, I_{R_1} is the current through resistor 1, I_{R_2} is the current through resistor 2, etc.

This formula was devised by Kirchhoff and is referred to as **Kirchhoff's current law.** It can be stated: "Whatever current flows into a circuit junction (or connection), must flow out." In the series circuit, whatever level of current enters one resistor, that same current level must flow out. In turn, the current enters the next resistor, flows out to the next resistor and so on.

History Hit!

Gustav Robert Kirchhoff (1824–1887).

Kirchhoff was educated in Russia, but spent much of his professional life in German institutions of higher learning. He is known for pioneering efforts in electrical network theory. He and chemist Robert Wilhelm Bunsen developed the spectrometer, a device that can analyze the makeup of various materials. It is also interesting to note that Kirchhoff was confined to a wheelchair due to an accident early in his life.

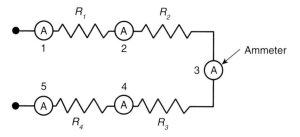

Figure 6-3.
All of the ammeters in the circuit will read the same current value.

In **Figure 6-3,** the circuit of Figure 6-2 has been redrawn to include ammeters to measure current. Every meter will read the same, regardless of the value of any one of the resistances. Of course, if any resistor is changed in value, the total current in the circuit will also change. Ohm's law shows this to be true.

$$I_T = \frac{E_T}{R_T}$$

Voltage Drop

In our study of the volt, we learned that it was the unit of measurement for potential difference or electromotive force (voltage). A force is required to cause a current to flow through a circuit. In **Figure 6-4,** the circuit has been redrawn and values assigned.

The total resistance of this circuit is equal to:
$R_T = 100\ \Omega + 100\ \Omega + 100\ \Omega + 100\ \Omega = 400\ \Omega$

The current flowing in all parts of the circuit is:
$$I = \frac{100\ V}{400\ \Omega} = 0.25\ \text{amps}$$

But point A has a potential of 100 volts and point B is zero. Has the 100 volts been lost somewhere along the circuit between A and B? You can understand how this can happen when you think of a resistance as something that must be overcome by a force. Actually the resistance has been overcome, but some of the force or voltage has been lost. The voltage lost by resistance in a circuit is called the *voltage drop,* or *IR drop.* It is called IR drop because Ohm's law states that:
$$E = IR$$

To find the voltage drop for each resistor is relatively simple. First find the current, which we have computed as 0.25 amps, then:
$$E_{R_1} = 0.25\ A \times 100\ \Omega = 25\ V$$
$$E_{R_2} = 0.25\ A \times 100\ \Omega = 25\ V$$
$$E_{R_3} = 0.25\ A \times 100\ \Omega = 25\ V$$
$$E_{R_4} = 0.25\ A \times 100\ \Omega = 25\ V$$

The voltage lost across R_1 is 25 volts, so the voltage at point C is 100 V – 25 V or 75 V.

The voltage lost across R_2 is 25 volts, so the voltage at point D is 75 V – 25 V or 50 V.

The voltage lost across R_3 is 25 volts, so the voltage at point E is 50 V – 25 V or 25 V.

The voltage lost across R_4 is 25 volts, so the voltage at point B is 25 V – 25 V or 0 volts.

Add up all the voltage drops and they will total 100 volts, which is the applied voltage of the circuit.

25 V + 25 V + 25 V + 25 V = 100 volts

In addition to his current law, Kirchhoff also formulated a voltage law that bears his name. *Kirchhoff's voltage law* states: "The sum of the voltage drops in a circuit must equal the applied, or source, voltage." Or:

$$E_T = E_{R_1} + E_{R_2} + E_{R_3} + E_{R_4}...$$

where E_T is the total applied voltage in the circuit, E_{R_1} is the voltage drop across resistor 1, E_{R_2} is the voltage drop across resistor 2, etc.

In **Figure 6-5,** we know the total current (0.5 amps) and the total resistance

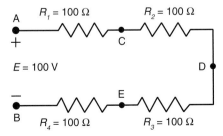

Figure 6-4.
The resistance and voltage are known. What are the voltage drops across each resistor?

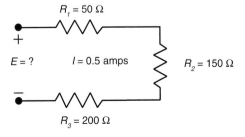

Figure 6-5.
The resistance and current are known. What is the value of the voltage applied to the circuit?

$(R_1 + R_2 + R_3 = R_T$ or $50\ \Omega + 150\ \Omega + 200\ \Omega = 400\ \Omega)$, but we do not know the applied voltage. Ohm's law states that:

$$E = IR$$

so,

$$E = 0.5\ A \times 400\ \Omega = 200\ V$$

The voltage drops can be easily found:

$$E_{R_1} = 0.5\ A \times 50\ \Omega = 25\ V$$
$$E_{R_2} = 0.5\ A \times 150\ \Omega = 75\ V$$
$$E_{R_3} = 0.5\ A \times 200\ \Omega = 100\ V$$
$$\text{Total} = 200\ V$$

The sum of the voltage drops is 200 volts. It is the same as we computed for the applied (source) voltage.

One more example shows another way that a problem can be presented. In **Figure 6-6,** we know the applied voltage (100 V) and the voltage drops across the resistors, but we do not know the value of the resistors. The current is 100 mA (0.1 amp).

According to Ohm's law:

$$R = \frac{E}{I}$$

The voltage drop across R_1 is 50 volts, so:

$$R_1 = \frac{50\ V}{0.1\ A} = 500\ \Omega$$

The voltage applied to R_2 is 25 volts, so:

$$R_2 = \frac{25\ V}{0.1\ A} = 250\ \Omega$$

The voltage applied to R_3 is 25 volts, so:

$$R_3 = \frac{25\ V}{0.1\ A} = 250\ \Omega$$

The problem can be proved by working it backwards. The total resistance equals $R_1 + R_2 + R_3$ ($500\ \Omega + 250\ \Omega + 250\ \Omega$) equals $1000\ \Omega$, or $1\ k\Omega$. Now,

$$I_T = \frac{E_T}{R_T} \text{ or } I = \frac{100\ V}{1000\ \Omega} = 0.1\ A$$

which proves that we did our calculations correctly.

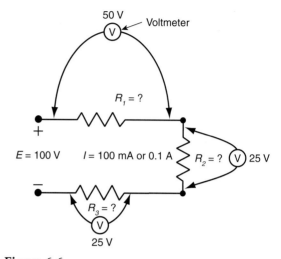

Figure 6-6.
The voltage drops and current are known. What are the values of the resistors?

There are thousands of applications of series circuits and voltage drops in electrical and electronic circuits. Time used to practice these problems and formulas will be time well spent.

Summary of the Laws of a Series Circuit

1. The total resistance in a series circuit is equal to the sum of all the individual resistors. ($R_T = R_1 + R_2 + R_3...$)
2. In a series circuit, the current flowing in all parts of the circuit is the same. ($I_T = I_{R_1} = I_{R_2} = I_{R_3}...$). This formula is known as Kirchhoff's current law.
3. The sum of the voltage drops in a series circuit is equal to the applied voltage. ($E_T = E_{R_1} + E_{R_2} + E_{R_3}...$). This formula is known as Kirchhoff's voltage law.

Practical Applications:
Project 1—Experimenter, can be found in Chapter 19. Conduct the experiments labeled, Problem 2 and Problem 3. Problem 2 involved the use of only one lamp.

Problem 3 uses three lamps wired in series. The lights in Problem 3 will glow much more dimly than the single lamp used in Problem 2.

Why is this so? The answer can be found in Ohm's law. Each light is a resistance unit. When three lights are in series, consider the sum of the resistances. Increased resistance will decrease the current.

Quiz–Chapter 6

Write your answers to these questions on a separate sheet of paper. Do *not* write in this book.

In the following problems:
I = current
R = resistance
E = applied voltage
E_{R_1} = voltage drop across R_1
E_{R_2} = voltage drop across R_2, etc.

Solve for the unknowns.

1.

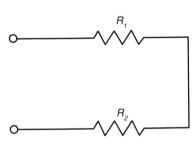

E = 50 V
R_1 = 50 Ω
R_2 = 150 Ω

R_T = _____
I_T = _____
E_{R_1} = _____
E_{R_2} = _____

2.

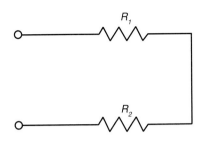

E_{R_1} = 25 V
E_{R_2} = 175 V
I_T = 0.5 A

E_T = _____
R_1 = _____
R_2 = _____
R_T = _____

3.

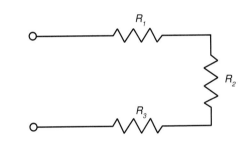

R_1 = 100 Ω
R_2 = 200 Ω
I_T = 2 A
E_{R_3} = 200 V

E_T = _____
R_3 = _____
R_T = _____
E_{R_1} = _____
E_{R_2} = _____

4.

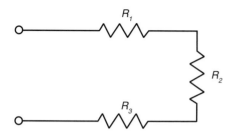

$E_{R_1} = 50$ V
$E_{R_2} = 150$ V
$E_{R_3} = 200$ V
$I_T = 0.25$ A

$E_T = $ _____
$R_1 = $ _____
$R_2 = $ _____
$R_3 = $ _____

Safety Suggestion!

Never use a lamp or appliance cord in your home if the insulation covering is worn out! If insulation from the cord is damaged, unplug the device and have the cord replaced. Never remove a plug from an outlet by pulling on the cord. Always pull by the plug. This prevents the cord from separating from the plug.

5.

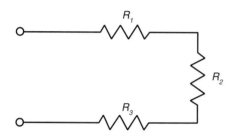

$E_S = E$ supply, source or applied
$R_1 = 1000$ Ω
$R_3 = 500$ Ω
$E_{R_2} = 10$ V
$I_{R_1} = 0.001$ A

$E_{R_1} = $ _____
$E_S = $ _____
$R_T = $ _____
$E_{R_3} = $ _____
$R_2 = $ _____

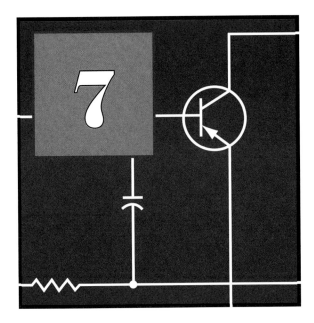

Parallel Circuits

Objectives

After studying this chapter, you will be able to answer these questions:
1. How are electrical components connected in parallel?
2. What are the laws of parallel circuits?

Important Words and Terms

The following words and terms are key concepts in this chapter. Look for them as you read this chapter.

conductance (G)

equivalent resistance

parallel

parallel circuit

siemens (S)

When we were studying the series circuit, the example of the highway between two towns was used. We learned that the several bridges along that highway were in series and each bridge offered a definite amount of resistance to the flow of traffic. As more automobiles are used each day, the need exists to provide better and faster highways. Your state may be constructing new or widened highways so more cars may travel safely without the resistance of narrow bridges and curves. The reason is apparent; more cars can travel on a new two- or three-lane roadway than could travel on the old one-lane road. In fact, twice as many cars can travel on a double road, three times as many on a three-lane road and so on.

If road A has the capacity of ten cars per minute and road B has a capacity of ten cars per minute, then roads A and B together have a capacity of 20 cars per minute, **Figure 7-1.** The bridges on each

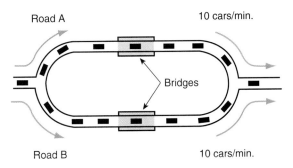

Figure 7-1.
The bridges are in parallel.

road may still offer some resistance to the flow of automobiles.

One could say that these roads are *parallel* to each other; that the bridge on road A is parallel to the bridge on road B. This same example can be applied to an electrical circuit. This example will help the beginning student in electricity to understand that in a parallel circuit the resistance decreases because more paths are provided for the flow of electricity.

Equal Resistors in Parallel

In **Figure 7-2**, the highways have been replaced by an electrical circuit. R_1 and R_2 are the bridges (resistance). The automobiles are replaced by electrons flowing along their highways, which are called conductors. At point X, the electrons divide, part taking road A and the other part taking road B. At point Y, the electrons rejoin and continue on their way. Thus, a *parallel circuit* has more than one path for current to flow. The parallel combination of resistance units offers less resistance to current than either single resistor.

In **Figure 7-3**, assume that R_1 equals 100 Ω and R_2 also equals 100 Ω. By combining the two in a parallel circuit, the total resistance of the circuit is only 50 Ω. The formula for finding the total resistance of a circuit (R_T) when the resistors are in parallel and all of the same value is:

$$R_T = \frac{R}{N}$$

where R_T is the total resistance, R is the value of any one resistor and N is the number of resistors in parallel.

Apply this formula to **Figure 7-3**,

$$R_T = \frac{100\ \Omega}{2} = 50\ \Omega$$

Like all circuits, there is a voltage, E, across the input terminals. This potential difference (voltage) causes the electrons to flow in the circuit.

Let's draw some conclusions about this circuit.

1. The voltage across all branches or paths of a parallel circuit is the same. In this case, it is the same as the applied voltage (E). You can see that the voltage across R_1 is the same as the voltage across R_2 since the ends of the resistors are connected to the common points X and Y that in turn are connected directly to the power source:

$$E_T = E_{R_1} = E_{R_2}$$

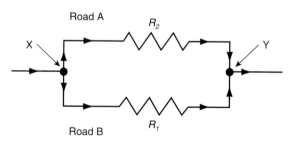

Figure 7-2.
These two resistors are in parallel.

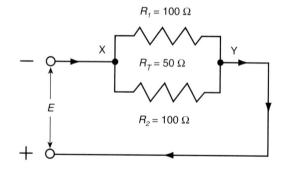

Figure 7-3.
Equal resistances in parallel.

Kirchhoff's voltage law is still true. That is to say, since all components are connected in parallel, there is only one voltage drop. That one voltage drop is equal to the source or applied voltage.
2. The total current in the circuit is equal to the sum of all the currents flowing in the branches of the parallel circuit or:

$$I_T = I_{R_1} + I_{R_2}$$

An example using actual values will help you understand this.

In **Figure 7-4**, total resistance (R_T) of the circuit can be found using:

$$R_T = \frac{R}{N} = \frac{200\ \Omega}{2} = 100\ \Omega$$

The current flowing in the circuit can be found by Ohm's law.

$$I = \frac{E}{R} = \frac{200\ V}{100\ \Omega} = 2\ A$$

When the current of 2 amperes reaches point X, it divides. One ampere flows around branch A through R_1, and one ampere flows around branch B through R_2. At point Y, the two currents rejoin.

The total current I_T = 2 amps
Current in branch A = 1 amp
Current in branch B = 1 amp
I_T = 1 amp + 1 amp = 2 amps

Again, note that Kirchhoff's other law, Kirchhoff's current law, still holds true. All current that flows into a junction will flow out of it.

Unequal Resistors in Parallel

So far we have assumed that R_1 equals R_2. This is not always so. Often they are unequal, and there is not an equal division of currents flowing in the branches of the parallel circuit.

In **Figure 7-5**, when the current reaches point X, it will still divide, but the greater amount of current will flow through the branch B, because branch B has less resistance than branch A.

Computing the total resistance of a parallel circuit having two unequal resistors is not too difficult. The following formula is called the *product over the sum method* due to the arithmetic operations involved. It can be used for any two resistors in parallel.

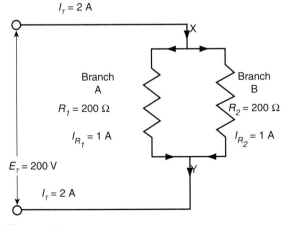

Figure 7-4.
The circuit current divides between the branches of the parallel circuit. Kirchhoff's current law still holds true!

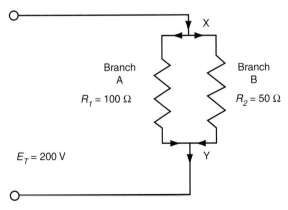

Figure 7-5.
Unequal resistances in parallel.

$$R_T = \frac{R_1 \times R_2}{R_1 + R_2}$$

In our circuit:

$$R_T = \frac{100\ \Omega \times 50\ \Omega}{100\ \Omega + 50\ \Omega} = \frac{5000\ \Omega}{150\ \Omega} = 33.33\ \Omega$$

The total current I_T is then:

$$I_T = \frac{200\ \text{V}}{33.33\ \Omega} = 6\ \text{A}$$

To find how much current will flow in branch A:

$$I_A = \frac{200\ \text{V}}{100\ \Omega} = 2\ \text{A}$$

In branch B:

$$I_B = \frac{200\ \text{V}}{50\ \Omega} = 4\ \text{A}$$

The sum of the branch currents ($I_A + I_B$) is 2 amps + 4 amps = 6 amps. This is the same as the total current flowing in the circuit. This again confirms Kirchhoff's current law.

Conductance

The ability to conduct electricity is opposite the ability to resist the flow of electricity. So we can consider the current carrying ability of any wire or circuit either by stating its resistance to the flow of electrons, or by its ability to conduct electrons. Its ability to conduct is called *conductance*. The letter symbol for conductance is *G*. Conductance is measured in *siemens (S)*. Conductance is the *reciprocal* of resistance. That is to say, conductance is one divided by the resistance value. For example, if the resistance of a circuit is 4 Ω, its conductance can be found using:

$$G = \frac{1}{R}$$

so

$$\frac{1}{4\ \Omega} = 0.25\ \text{siemens (S)}$$

History Hit!

Ernst Werner von Siemens (1816–1892)

Siemens is honored with the unit of conductance being named after him. Educated as an electrical engineer in Germany, he led the way for many advances in the principles of electricity. For example, he developed a method to coat wire with an insulation that was seamless—quite an advance for his time. He founded a German company that still carries his name and continues its founder's refinements in the field of electrical devices and generating equipment.

If the circuit resistance is 500 Ω, its conductance is:

$$G = \frac{1}{R} = \frac{1}{500\ \Omega} = 0.002\ \text{S}$$

Two or More Resistors in Parallel

When two or more unequal resistors are connected in parallel, the conductance method of finding R_T is:

$$R_T = \frac{1}{\frac{1}{R_1} + \frac{1}{R_2} + \frac{1}{R_3}}$$

This formula involves the use of fractions, and is sometimes called the *reciprocal method*. An example will show you how it works. See **Figure 7-6**.

$$R_T = \frac{1}{\frac{1}{100\ \Omega} + \frac{1}{200\ \Omega} + \frac{1}{400\ \Omega}}$$

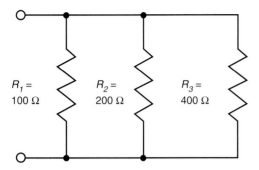

Figure 7-6.
Three unequal resistors in parallel.

Find the least common denominator of the fraction, which is 400.

$$R_T = \frac{1}{\frac{4}{400} + \frac{2}{400} + \frac{1}{400}}$$

$$R_T = \frac{1}{\frac{7}{400}}$$

now,

$1 \div \frac{7}{400}$ is the same as

$$1 \times \frac{400}{7} = \frac{400}{7} = 57.1 \ \Omega$$

$$R_T = 57.1 \ \Omega$$

Another technique is to simply divide each fraction into one, add the values, then divide into one again.

$$R_T = \frac{1}{\frac{1}{100 \ \Omega} + \frac{1}{200 \ \Omega} + \frac{1}{400 \ \Omega}}$$

$$R_T = \frac{1}{0.01 \ S + 0.005 \ S + 0.0025 \ S}$$

$$R_T = \frac{1}{0.0175 \ S}$$

$$R_T = 57.1 \ \Omega$$

Remember to be careful when adding decimal numbers. All decimal points must line up to be added together properly:

$$\begin{array}{r} 0.01 \\ 0.005 \\ +\ \ 0.0025 \\ \hline \text{Sum} = 0.0175 \end{array}$$

Note: In all problems dealing with resistance in parallel circuits, the total resistance must always be *less* than the value of any resistor in the parallel circuit. Recall that parallel circuits provide additional pathways for current flow. Total resistance always decreases the more parallel resistances are added to the circuit. Use this information to check your work.

Equivalent Resistance

The flow of electricity in a circuit depends upon resistance of the circuit. This resistance can be a single resistor or several resistors connected in series or parallel. Regardless of how many resistors there are or how they are connected, they will combine together to give a total resistance in the circuit. The total resistance is the limiting factor affecting the current. In other words, the total of all resistances might be represented by one resistor value. This resistance is called the *equivalent resistance* of the circuit. See **Figure 7-7.**

Step I. Combine R_2 and R_3

$$R_T \ (R_2 \text{ and } R_3) = \frac{400 \ \Omega \times 100 \ \Omega}{400 \ \Omega + 100 \ \Omega}$$

$$R_T = \frac{40{,}000 \ \Omega}{500 \ \Omega} = 80 \ \Omega$$

The circuit will appear as shown in **Figure 7-8.**

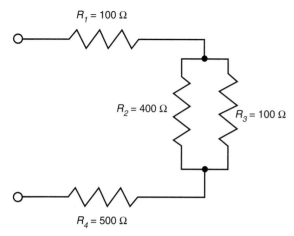

Figure 7-7.
A combination circuit of series and parallel resistors.

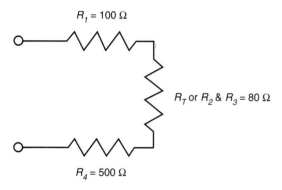

Figure 7-8.
Resistors R_2 and R_3 have been combined.

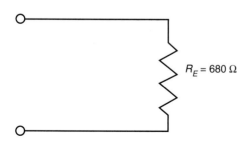

Figure 7-9.
The equivalent resistance of the series and parallel combination circuit.

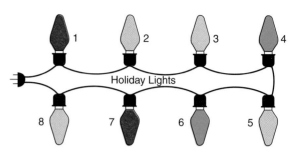

Figure 7-10.
Holiday lights wired in series.

Step II. Combine the three resistors in Figure 7-8. They are in series so:

$$R_1 + (R_2 \text{ and } R_3) + R_4 =$$
$$100 \text{ } \Omega + 80 \text{ } \Omega + 500 \text{ } \Omega = 680 \text{ } \Omega$$

The circuit can now be represented by **Figure 7-9**.

Electrically speaking, the circuits of Figures 7-7, 7-8, and 7-9 are exactly the same. R_E (680 Ω) is the equivalent resistance of the combination of R_1, R_2, R_3, and R_4.

Applications

Before leaving the study of series and parallel circuits, let's look at some familiar applications.

A string of holiday or party lights could be connected in a series or parallel manner. Compare **Figures 7-10** and **7-11**.

The symbol ⊕ is used for a lightbulb.

In Figure 7-10, the eight lightbulbs are connected in series. All electrical current in the circuit must pass through each lightbulb.

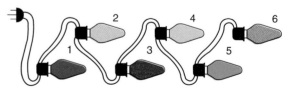

Figure 7-11.
Holiday lights wired in parallel.

Chapter 7   Parallel Circuits

Safety Suggestion!

Remember the voltage and current levels present at all electrical connections in your home can be lethal. This means you need to take special safety precautions when exposed to such circuits. The safest action step you can take is to de-energize the circuit you are examining. This is done by removing the fuse or switching off the circuit breaker to the circuit. The electrical voltage (or pressure) coming from a household outlet is sufficient to push enough current through your body to cause death.

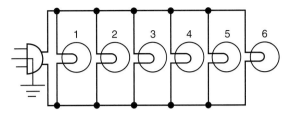

Figure 7-13.
Schematic drawing for holiday lights connected in parallel. Again, note the safety ground symbol found at the plug.

If one lightbulb should burn out, all the lights would go out.

Figure 7-12 shows exactly the same circuit as in Figure 7-10 except that conventional schematic symbols are used. This is the way you would find the circuit in the usual electrical drawing.

In Figure 7-11, the same lights are connected in parallel. The current divides between the six branches of the circuit. If one light should burn out, the remaining five will still operate properly.

Figure 7-13 shows the schematic drawing for the six holiday lights connected in parallel.

The electrician, in wiring your home, wired all your lights, convenience outlets, and appliances in parallel. We will now apply our knowledge of parallel circuits. See **Figure 7-14**.

Here we have two convenience (duplex) outlets or receptacles wired in parallel across the line voltage. The 110 volt line is made up of two electrical conductors and a safety ground. The conductors are referred to as the hot and neutral wires. The hot wire, usually protected by black (sometimes red) insulation provides the source of the electrical energy. This "hot"

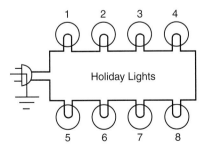

Figure 7-12.
Holiday lights in series using conventional schematic symbols. Note the electrical safety ground symbol found at the plug.

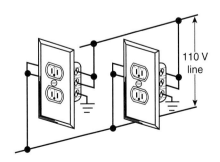

Figure 7-14.
Convenience (duplex) outlets in your home are connected in parallel across the line.

conductor is wired to the brass colored screw terminal of the duplex outlet.

The neutral (sometimes called the cold) wire is usually covered with white insulation and completes the circuit to the device plugged in. The neutral is termed this because it will ultimately be connected to an earth referenced ground. This ground connection can be the cold water pipe (as long as it is metal) or a copper rod pounded into the ground (earth) near the home electrical panel. The neutral wire is connected to the silver colored screw terminal on the duplex outlet. If you look carefully at the front of an outlet, the neutral wire is connected to the larger opening, the hot wire to the smaller opening.

The safety ground wire, or ground, is covered with green insulation. Its purpose is to conduct electrical current away from a fault or short circuit. Let's suppose an electrical heater that is made of metal develops a problem or fault right where the electric cord enters the device. The insulation has been damaged, and the hot wire is touching the metal cabinet. In this unsafe situation, someone touching the metal cabinet of the heater and an electrical ground (another appliance, plumbing, or any other device that may be grounded) could receive a fatal electrical shock. The safety ground takes this potential path of current (from the metal case) to the earth ground. The ground wire provides a low resistance connection from the hot wire to ground. In this way, an excessive amount of current flows. Enough current should flow that it will blow out a fuse or trip a circuit breaker. For this system to function properly, those who do the wiring, electricians, must follow strict codes and rules for everyone's safety. The National Electrical Code specifies the rules that must be followed by electricians who work in homes and industries.

In Figure 7-14, no electric current is being used because no appliance or light (an electrical load) has been plugged in.

When a load is plugged in ($R_1 = 100 \, \Omega$), a current flows, **Figure 7-15**.

$$I = \frac{110 \text{ V}}{100 \, \Omega} = 1.1 \text{ A}$$

and the power is:

$$P = I \times E$$

or

$$P = 1.1 \text{ A} \times 110 \text{ V} = 121 \text{ W}$$

R_1 (100 Ω) is the only resistance in the circuit.

When R_2 (also 100 Ω) is plugged in, R_1 and R_2 are in parallel and the total resistance is:

$$R_T = \frac{100 \, \Omega}{2} = 50 \, \Omega$$

and the total current flowing is:

$$I = \frac{110 \text{ V}}{50 \, \Omega} = 2.2 \text{ A}$$

The power is:

$$P = 2.2 \text{ A} \times 110 \text{ V} = 242 \text{ W}$$

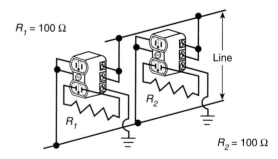

Figure 7-15.
Appliances plugged into convenience outlets are in parallel across the line.

Web Wanderings!

http://www.electronic-circuits-diagrams.com/

The Electronics Zone contains free electronic circuit diagrams with a complete explanation of how the circuit works. Students and hobbyists enjoy reviewing the wide range of circuits listed on this site, as well as reading the free tutorials in basic electronics.

You can now see that as you plug in more and more appliances, lamps, or other electrical loads, the current increases. If current increases, with the voltage at a fixed value (approximately 110 volts), the wattage consumed by these loads is also increased. In this way, your electrical bill also increases. In order to conserve energy and save money, always turn off any unnecessary load, be it the radio, television, or light.

Refer again to the *Project 1—Experimenter* in Chapter 19 of this textbook. In Problem 4, the lights are connected in parallel. Each light burns at the same brilliance as the single light of Problem 2. Because the circuit is a parallel circuit, each light decreases the total resistance of the circuit. The current, of course, will increase (Ohm's law).

Quiz–Chapter 7

Write your answers to these questions on a separate sheet of paper. Do *not* write in this book.

1. Write the formula for total resistance when all resistors in a parallel circuit are equal.
   $R_T =$ _____.
2. Write the formula for total resistance of two unequal resistors in parallel.
   $R_T =$ _____.
3. Write the formula for total resistance of three unequal resistors in parallel.
   $R_T =$ _____.
4. The sum of the branch currents in a parallel circuit is equal to the _____ of the circuit.
5. The voltage across all branches of a parallel circuit is _____.
6. Conductance has the letter symbol ____ and is measured in ____. It is the ____ of resistance.
7. What is the conductance of a circuit that has ten ohms resistance?
8. Find:
   $R_T =$ _____
   $I_T =$ _____
   $I_{R_1} =$ _____
   $I_{R_2} =$ _____

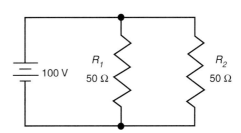

9. Find:
   $R_T =$ _____
   $I_T =$ _____
   $I_{R_1} =$ _____
   $I_{R_2} =$ _____

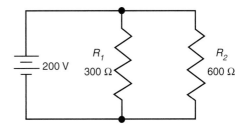

10. Find:
    $R_T = $ _____
    $I_T = $ _____
    $I_{R_1} = $ _____
    $I_{R_2} = $ _____
    $I_{R_3} = $ _____

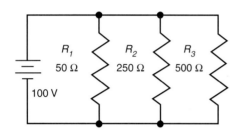

11. Find:
    $R_T = $ _____
    $E_T = $ _____

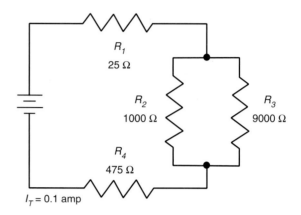

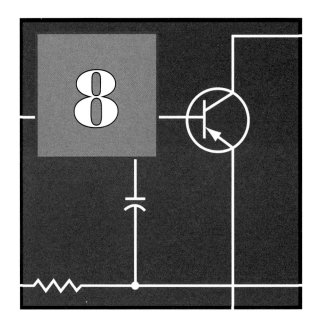

Sources of Electricity– Batteries

Objectives

After studying this chapter, you will be able to answer these questions:
1. What produces electricity?
2. How are batteries constructed and used?

Important Words and Terms

The following words and terms are key concepts in this chapter. Look for them as you read this chapter.

ampere-hour	hydrometer
battery	local action
cell	primary cell
dry cell	secondary cell
electrode	specific gravity
electrolyte	voltaic cell
electroscope	

In previous lessons you learned about the unit of electrical pressure, the volt. This voltage is caused by *a difference in potential* between two points. This difference in potential was necessary to cause an electric current to flow through a circuit. In this chapter we will discuss the methods of producing electricity.

Sources of Electricity

There are six methods of producing a difference in potential or a voltage. These six methods are commonly known as *sources of electricity.*

1. Chemical action — cell or battery
2. Magnetism — generator
3. Friction — static electric charge
4. Heat — thermocouple
5. Light — photovoltaic or solar cell
6. Pressure — crystalline materials (quartz)

The Voltaic Cell

In the middle of the eighteenth century, the Italian physicist Alessandro Volta conducted experiments that led to the discovery of the *voltaic cell.* The words voltaic and volt were named in memory of this great scientist.

History Hit!

Alessandro Volta (1745–1827)

Volta was from an aristocratic family. He was professor of natural philosophy (what we would call science today) and a highly religious individual. Experiments showed Volta that electricity could be created by bringing different metals into contact with each other. His cell, or "battery," was made of silver and zinc disks with a saltwater soaked card between them. A series of disks provided a steady electric current.

If we take a piece of zinc and suspend it in a glass dish filled with acid, we find the acid chemically reacts with the zinc. Little bubbles of hydrogen gas accompany this action, as the zinc is eaten away by the acid. Were we to electrically test this piece of zinc, we would find that it had a negative charge. The instrument used to make this test is called an *electroscope.*

Now, suspend a carbon rod in the same acid solution with the piece of zinc. A test of the carbon rod reveals that it is positively charged. A *potential difference* has been created. If a wire is connected between the carbon rod and the zinc plate, a current will flow. It continues to flow as long as there is chemical action.

In **Figure 8-1**, the carbon and zinc elements are called **electrodes** and the acid solution is known as the *electrolyte.* Many different kinds of materials and acids have been used in experimental voltaic cells.

Two defects of a carbon–zinc cell should be considered. The zinc plate contains some particles of carbon. Coal and coke are used in the smelting, or melting, process of extracting the zinc from the ore. Both coal and coke (made from coal) are made from carbon based materials. These particles of carbon are released as the zinc is eaten away. They become positively charged in the same manner as the carbon electrode. They attach themselves to the zinc electrode and form small voltaic cells within the larger voltaic cell. These little cells contribute nothing to the energy that can be supplied by the cell. This internal chemical action creating worthless energy is called **local action.**

The hydrogen bubbles that are formed when the zinc plate is placed in the acid are attached to the positive carbon electrode. These bubbles form a coating around the carbon and effectively insulate it from the chemical reaction. The voltage of the cell drops, as a result of this undesirable action. The cell is said to be *polarized.* Various compounds are used in cells that unite with the free hydrogen and counteract this polarization.

Cell Connections

The terms *cell* and *battery* are used in discussions throughout this chapter. Actually, a single unit is a **cell.** When several cells are connected together, they are known as a **battery.** In practice, the term battery is used rather loosely, and it usually means either a cell or a battery.

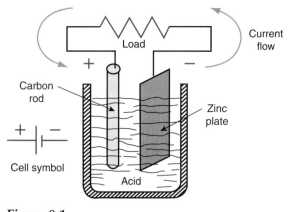

Figure 8-1.
A simple voltaic cell.

The terminal voltage of a lead–acid cell is about two volts. In order to get a voltage of six or twelve volts, such as required for an automotive battery, three or six cells respectively are connected together and placed in one container.

Primary Cells

Primary cells are by definition non-rechargeable cells. There are many types of primary cells in use today. Among the more common primary cells are: carbon–zinc, mercury, and alkaline cells.

Carbon–Zinc

The liquid voltaic cell has limited use today. Another version of the same cell is the *dry cell*. These are used extensively in flashlights, CD players, and portable radios. This dry cell consists of a zinc container, in the center of which is a suspended carbon rod. Study the construction shown in **Figure 8-2**.

The container is filled with a moist paste of manganese dioxide, carbon, and electrolyte. The double chemical reaction during discharge of the cell produces free electrons that make the zinc container the negative terminal and the center carbon electrode positive.

A cell constructed in this manner creates a potential difference (voltage) of about 1.5 volts. After the chemical action has been used up, the cell is dead.

Mercury Cells

Mercury cells are designed to provide high energy and stable voltage. They are used in electronic products ranging from watches to smoke alarms to cardiac pacemakers. In addition, their stable voltage enables them to be used as a voltage reference or standard. Their high energy output permits extensive miniaturization of products that utilize these cells.

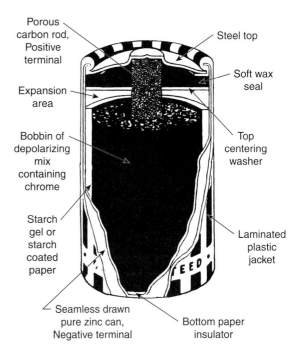

Figure 8-2.
A cutaway view of a dry cell. (Burgess Battery Co.)

Mercury cells are available in sizes ranging from 0.005 to 3.0 cubic inches. The basic chemicals used in their production are a mercury oxide cathode and a zinc anode (compacted powders) and an alkaline electrolyte (liquid).

These cells are made in two general types:
1. Pure mercuric oxide.
2. Mercuric oxide with a small fraction of manganese dioxide.

The first type, at 1.35 volts, maintains extremely stable voltage over a considerable period of time. It is used as a voltage reference and in high reliability applications. The second type of mercury cell, at 1.40 volts, is used in general applications where the need for stable voltage is less critical.

Mercury cells are being used less and less because of the toxic material used to make them. Mercury is a very dangerous material, and extreme caution must be taken with these cells if they leak their electrolyte. Other specialty cells can be used to replace the mercury cell.

Mercury cells come in three basic constructions:
1. Flat cells used in watches.
2. Cylindrical cells found in general use.
3. Wound anode cells that feature improved low temperature performance.

Alkaline Cells

Alkaline manganese cells are used to power many products where the voltage stability and high energy of the mercury system are not critical. Alkaline cells are used in all types of electrical/electronic consumer devices, both in original equipment and in the replacement market.

The basic components of the alkaline system are a manganese dioxide cathode, a zinc anode, and an alkaline electrolyte. As with mercury cells, the cathode and anode are compacted powders, while the electrolyte is a paste. These components are sealed in a steel jacket with a plastic top. End caps and outer steel jackets complete the assembly.

Alkaline manganese cells are available in cylindrical and button shapes. The cylindrical cells are in common use in many consumer devices. The flat cells are used in products that require a compact power source. These cells are capable of leaking electrolyte when depleted.

Secondary Cells

Secondary cells are known as rechargeable cells. These types of cells can be recharged many times, which increases their usefulness. Common secondary cells are: nickel–cadmium, nickel-metal hydride, lithium ion, and the lead-acid storage cell.

Nickel–Cadmium Cells

Nickel-cadmium (Ni-Cd) cells are used in many types of cordless home appliances such as razors, radios, small power tools, and flashlights. They are recharged by plugging them into the customary home power outlet. These cells have a low internal resistance and deliver high currents with little loss of terminal voltage.

Any description of the Ni-Cd cell would be quite technical. However, the plates or electrodes are made by sintering (heating) powdered nickel into a nickel wire screen. This makes a strong and flexible plate. The positive plates are impregnated with nickel salt solutions. Cadmium salt solutions are used for the negative plates. The electrolyte is a solution of potassium hydroxide. The terminal voltage of the cell is about 1.33 volts. These cells are being used in fewer applications because of the toxic characteristics of the materials inside them.

Nickel-Metal Hydride Cells

Nickel-metal hydride (Ni-MH) cells have often been used in laptop computers, audio/visual equipment, and cordless appliances. Nickel-metal hydride cells have a greater power capacity than nickel-cadmium cells, and they last longer. One very important characteristic of the nickel-metal hydride cell is that it does not contain cadmium. Cadmium is a toxic material that must be handled carefully. This is why nickel-cadmium cells are not as popular as they once were. Nickel-metal hydride cells, however, do not operate in as wide a temperature range as the nickel-cadmium.

Lithium Ion Cells

Lithium ion (Li-Ion) cells are becoming more widely used in laptop computers. They pack as much energy as the nickel-metal hydride cell, and they weigh less. Lithium ion cells are more expensive than nickel-metal hydride, but their lightweight characteristic makes them an attractive power source.

Lead-Acid Storage Battery

The lead-acid storage battery is typically called a car battery, because this is where this battery is most often found. From automobiles, lawn equipment, and marine applications, lead-acid storage batteries are very common. Actually a storage battery does not store up electricity. It would be more correct to say that it stores a chemical action that produces electricity. The lead-acid cell is said to be reversible. This means that it can be recharged and used over and over again.

The plates or electrodes of the lead-acid cell are made of chemicals that have been pressed into a grid-like frame. This frame acts as a rigid support and prevents the chemicals from falling apart. The negative plates are made of spongy lead and are slate gray in color. The positive plates are made of lead peroxide and are reddish brown. Each cell contains several of these thin plates. So that the maximum area of the plates can be used in the chemical action, the plates are interlaced, placing a positive plate between two negative plates. So that no plate will touch another plate, separators of wood, glass, or rubber are used. Additional materials are also used as separators. A thirteen-plate cell would have seven negative and six positive plates, **Figure 8-3.**

This assembly of plates and separators is placed in a hard rubber, glass, or plastic case with the necessary connections and terminals. Usually three or six cells (2 V/each) are placed in one case. The cells are filled with an electrolyte of sulfuric acid (H_2SO_4) and water.

The chemical actions during charge and discharge are illustrated in the drawings in **Figure 8-4.** Note, in particular, that during discharge the plates became

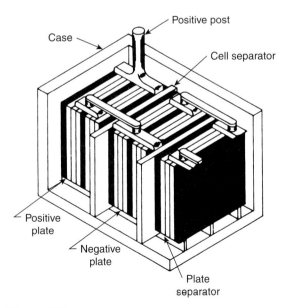

Figure 8-3.
The negative and positive plates are interlaced with separators between them.

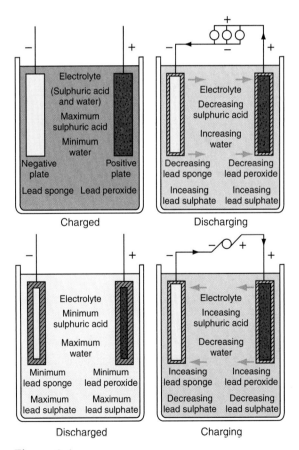

Figure 8-4.
The chemical action during charge and discharge is illustrated.

increasingly lead sulfate and the electrolyte becomes nearer to that of water. During charge, the plates approach their original state and the electrolyte is richer with acid.

Reversing the current through a battery returns the battery to its original state of charge. Battery chargers or generators perform this task, **Figure 8-5.** A charger is plugged into line (120 Vac) voltage. The transformer in the charger changes the voltage level needed for various types of batteries (6 volt or 12 volt systems). A rectifier (diode) then changes alternating current (ac) into pulsating direct current (dc) that can be used to recharge a battery.

Be sure all vent holes in the cabinet of the charger are clear because the transformer and rectifier units can heat up. Also, be sure the clips to connect on the battery do not touch. This would place a short circuit on the output of the charger. A fuse or circuit breaker should protect internal components from damage. Some chargers include a meter movement. The meter can be a voltmeter or ammeter. Some chargers use a switch to change the meter from reading voltage or current. Follow all manufacturers' directions related for safely

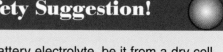

1. Battery electrolyte, be it from a dry cell (carbon-zinc) or a wet cell such as the lead-acid storage battery, is extremely dangerous. It will cause chemical burns on your skin, or it can damage your vision if the substance gets into your eyes. Additionally, the acid will attack and destroy the fibers that make up your clothing. Should battery acid get on your hands or skin, wash the acid off immediately. Should chemical burns be evident, seek first aid at once. If any acid reaches your eyes, flush them at once with water and seek medical attention as soon as possible to prevent further damage. Battery acid on your clothes can be neutralized by a solution of baking soda and water.
2. While charging, lead-acid storage batteries produce hydrogen gas. This is a by-product of the charging process. Hydrogen gas is extremely flammable. Never bring an open flame (such as a lit match or lighter) near a charging battery. As an added precaution, always charge a battery in a well-ventilated area. It is also a wise suggestion to wear eye protection when working around batteries.
3. Lead-acid batteries are made of a large quantity of lead, which is a very dense (heavy) material. Additionally, the electrolyte solution adds even more weight to the battery. When necessary, get assistance to move lead-acid storage batteries. They are heavier then you think. If possible, use carrying straps to make battery movement a bit easier.

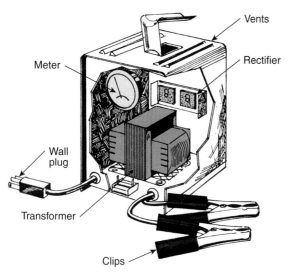

Figure 8-5.
This battery charger uses a transformer to reduce a 120 volt source to the voltage of the battery. (Triple A Specialty)

charging batteries. A simple battery charger circuit is found in Chapter 19, *Project 10—Automotive Battery Charger.*

An indication of the state of charge of a lead-acid cell is the amount of sulfuric acid in the electrolyte. All liquids have a specific weight in relation to an equal volume of water. This is known as the ***specific gravity*** of the liquid. The specific gravity of water is 1.000 and the specific gravity of sulfuric acid is 1.840. This means that sulfuric acid is 1.84 times heavier than water. A fully charged lead-acid cell will have a specific gravity of about 1.300. As the battery discharges, the acid is used up and water replaces it. This specific gravity decreases to between 1.100 and 1.150 for a fully discharged battery. The device used to measure the specific gravity of a liquid is called ***hydrometer.***

Maintenance Free Lead-Acid Storage Batteries

In the past, most automotive batteries needed to be replenished with pure, clean water added to the chemical electrolyte. Currently, many automotive batteries are "maintenance free" types. Maintenance free batteries are better sealed to prevent and reduce evaporation of the electrolyte and therefore, need little to no maintenance as to electrolyte level. Terminals and connections should be kept clean and tight, however, or battery life can be shortened.

Series and Parallel Connections

When cells are connected in series (positive to negative, positive to negative, and so on), the final voltage is equal to the voltage of one cell times the number of cells.

$$E_{one\ cell} \times N = E_{battery}$$

where N is the number of cells.

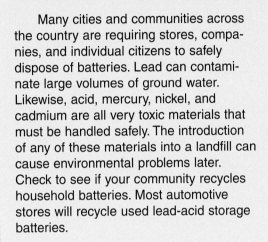

Environmental Update!

Many cities and communities across the country are requiring stores, companies, and individual citizens to safely dispose of batteries. Lead can contaminate large volumes of ground water. Likewise, acid, mercury, nickel, and cadmium are all very toxic materials that must be handled safely. The introduction of any of these materials into a landfill can cause environmental problems later. Check to see if your community recycles household batteries. Most automotive stores will recycle used lead-acid storage batteries.

Figure 8-6 shows three 1.5 volt cells connected in series. The battery voltage is 4.5 volts.

When cells are connected in parallel (positive to positive, and so on), the final battery voltage will equal only the voltage of one cell. However, you will have a much stronger battery, and it will last longer. This is because in parallel, the current from each cell joins together. The total current in the circuit equals the sum of each individual cell current flow.

Figure 8-7 shows three 1.5 volt cells connected in parallel. However, current flow can be increased up to three times. The

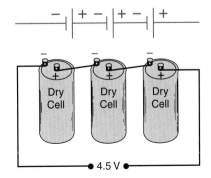

Figure 8-6.
Three cells connected in series.

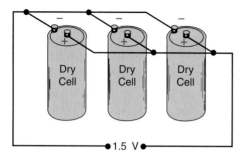

Figure 8-7.
Three cells connected in parallel.

battery voltage is 1.5 volts. Combinations can be connected for desired voltages and capacities. The symbol for a battery (two or more cells) in a diagram is:

Battery Capacity

The capacity of a battery is rated in *ampere-hours.* An automobile battery usually has a rating of 120 or greater ampere-hours depending upon the number of plates in the battery and size. The larger the engine that needs to be started, the greater the battery capacity that is needed.

A 100 ampere-hour battery theoretically will deliver,

1 amp for 100 hours

or

10 amps for 10 hours

or

any combination that will give a product of: amperes × hours = 100

This capacity is determined under controlled laboratory conditions. If a battery is discharged rapidly, it will not deliver to its full capacity. It should also be noted that temperature affects battery operation. Generally speaking, as temperature drops, so does the chemical action that takes place inside a cell or battery. This decrease in chemical action limits the output of a battery.

Quiz–Chapter 8

Write your answers to these questions on a separate sheet of paper. Do *not* write in this book.

1. A cell that cannot be recharged is called a(n) _____ cell.
2. A cell that can be recharged is called a(n) _____ cell.
3. In the lead-acid cell, the negative plates are made of _____ and are _____ in color.
4. In the lead-acid cell, the positive plates are made of _____ and are _____ in color.
5. The voltage of a lead-acid cell, is about _____ volts.
6. Two defects of the voltaic cell are _____ and _____.
7. Many cordless types of home appliances use _____ cells. These cells are _____ cells and can be _____.
8. Show in a diagram the connections that are required for six two-volt cells to produce a twelve-volt battery.
9. Show in a diagram the connections that are needed for three two-volt cells to produce a six-volt battery.
10. When charged, the specific gravity of a lead-acid cell is _____. When discharged, the specific gravity of a lead-acid cell is _____.
11. The capacity of a battery is rated in _____.
12. What should you do if some battery acid spilled on your hands and clothes?
13. Why will a discharged lead-acid battery freeze in cold weather?
14. A newspaper advertisement states that a certain special battery has 15 plates and is rated 120 ampere-hours. What is the meaning of these specifications in terms of battery capacity?

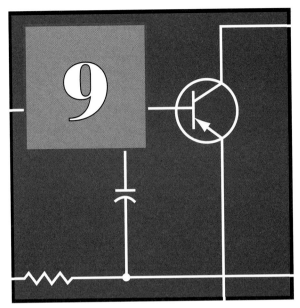

Sources of Electricity— Friction, Heat, Pressure, Light

Objective

After studying this chapter, you will be able to answer this question:
1. How can friction, heat, pressure, and light be converted to electrical energy?

Important Words and Terms

The following words and terms are key concepts in this chapter. Look for them as you read this chapter.

photocell

photoresisor

photovoltaic cell

piezoelectric effect

solar cell

static electric charge

thermocouple

There are two primary sources of electric power generation currently used. First, batteries, which use chemical action to create electricity, are found in a great many electronic devices. Cells and batteries are discussed in the previous chapter. Electric generators, using magnetism, are the second primary method employed to produce electricity. Generators are used to produce large quantities of electricity, which is used in residential and commercial purposes. Magnetic generators are discussed in Chapter 12.

However, there are several other methods of producing electricity. These methods are not used as commonly, but they are still important to understand. The methods are friction, heat, pressure, and light. They each have important applications.

Voltage from Friction

We learned in an early lesson that some atoms gave up their electrons when rubbed by a cloth. Remember the Greek "elektron?" Familiar examples of this phenomenon are demonstrated in the laboratory by rubbing a glass rod with a silk cloth. The glass rod becomes positively charged. Rub a piece of vulcanite (hard rubber) with fur and it becomes negatively charged. These charges are generally considered "electricity at rest" or *static electric charges*. However, the potential difference between the two can cause a flow of electric current. It is possible to build up very high static charges by means of friction.

The physics laboratory in your school may have a static electric generating machine. The Wimshurst generator, **Figure 9-1,** has two moving plastic disks. As they rotate, very thin and tiny wires (like those found in a wire brush) rub along metal sections on the disks. The developed charge is then transferred to capacitors or Leyden jars. When the charge is a sufficient size, a miniature bolt of lightning or spark jumps across the gap between the metal spheres.

Another type of static machine is a Van de Graaff generator, **Figure 9-2.** An electric charge is carried up into the sphere by means of a motor-driven belt and roller assembly, which are selected for their static electric properties. The charge is picked up from the belt and is collected on the outside of the metal sphere. Large static charges can be collected with this generator to produce "artificial lightning."

Electric motors that drive belts and pulley systems can also build up static

Safety Suggestion!

If you should use one of these machines, be sure to discharge the Leyden jars by bringing the two ball-type electrodes together. Otherwise, someone could get an unhealthy shock.

electric charges. Therefore, many motor and pulley systems are grounded for electrical safety reasons.

Lightning, one of the most destructive forces of nature, is a very convincing demonstration of electric charges. Rapidly moving air currents during a storm charge the clouds with static electricity. An enormous potential difference can be built up between two clouds or between a cloud

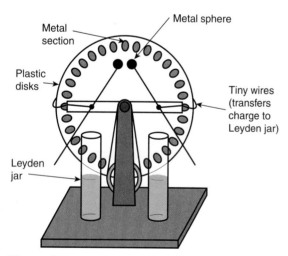

Figure 9-1.
A Wimshurst static generator.

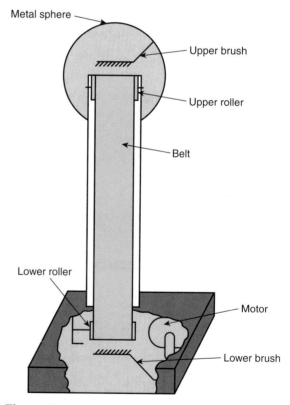

Figure 9-2.
A Van de Graaff generator.

Chapter 9   Sources of Electricity—Friction, Heat, Pressure, Light

> **Safety Suggestion!**
>
> Remember safety precautions to take when a thunderstorm approaches. Try to remain inside. Do not use the telephone if possible. Televisions with outside antennas or cable connections are also a source of possible lightning damage. It would be wise to turn off the television or video game until the storm passes. If you are outside, stay out of open fields and stay away from trees. Lightning usually strikes the highest object in the area. It is best to be very cautious during any lightning storm.

and the ground. When the charge is great enough, a great spark jumps across in the form of lightning, discharging the cloud.

Voltage from Heat

If two dissimilar pieces of metal are twisted together and heated, a potential difference or voltage develops across the ends of the wires. Such a device is known as a *thermocouple*, **Figure 9-3**.

Although the voltage is very small, it can be used to indicate the temperature of the heat. Commercially made thermocouples are used extensively as heat indicating and control devices. A meter attached to the ends of the thermocouple responds to a change in voltage. The dial on the meter can be calibrated to read in degrees of heat.

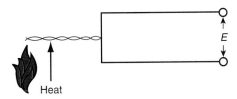

Figure 9-3.
A diagram of a simple thermocouple.

Voltage from Pressure

Certain natural crystalline substances, such as quartz, have very unusual properties. If a slab of crystalline quartz is placed between two metal plates and a pressure or vibration is applied, a voltage is developed between the two plates, **Figure 9-4**.

This unusual property of a quartz crystal is called the *piezoelectric effect.* This effect was a very useful discovery. Used extensively in the past with phonograph records, this crystal formed part of the pickup needle. These days, piezoelectric strikers are used with outdoor gas grills and some soldering and brazing torches to create the spark necessary to ignite the fuel gas.

Crystals are also used in much the same manner in a crystal microphone. The sound waves from voice or music cause the plates to exert a pressure on the crystal, which converts this pressure into electrical energy. The voltage is amplified by electronic circuits and used to operate a loudspeaker.

Voltage from Light

When light is directed onto a photosensitive material, electrons are freed from their orbits. This is the principle that provides the operation of the *photovoltaic cell,*

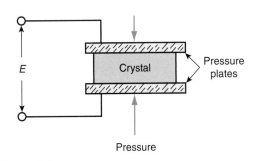

Figure 9-4.
Pressure exerted on the crystal by the contact plates creates a voltage.

or *solar cell*. Solar cells convert light energy into electrical energy. See **Figure 9-5.** Space vehicles and satellites use solar cells to provide the electrical energy needed to operate their instrumentation. A great deal of research has been completed to discover ways to convert the energy of the sun into useful power via this method.

A *photocell* is another photosensitive device. The electrical properties of a photocell are modified by light. It is used in science, engineering, and medicine for all kinds of delicate counting and grading (quality checking) applications. A camera light meter may use a photocell, which indicates the correct camera setting for the light intensity. Photocells can be connected in series to generate a sufficient voltage to operate a radio or other devices. Today most of this technology is built into digital cameras.

Another popular light-sensitive device is the *photoresistior.* This component changes its resistance under the influence of light intensity. A change of resistance in a circuit also changes the current in a circuit. The change in current can be indicated on a meter. A similar device can be used as a light meter by a photographer.

You will be interested to know that the photoelectric effect plays a major role at the television studio. In the television camera, a scene is focused upon a mosaic of many tiny photoelectric cells, which convert the different shades and intensities of the picture into voltages. These voltages are used by the transmitter to send the picture.

Figure 9-5.
Solar cells are often used to power small, low-power devices like this calculator.

Quiz–Chapter 9
Write your answers to these questions on a separate sheet of paper. Do *not* write in this book.
1. List six sources of electrical energy.
2. The property of a quartz crystal to generate a voltage when pressure is applied to it is called the _____ effect.
3. Two dissimilar metals can be joined together to measure temperature. This device is called a(n) _____.
4. Name two common applications of quartz crystals in electric circuits.
5. List six ways that a photocell might be used.

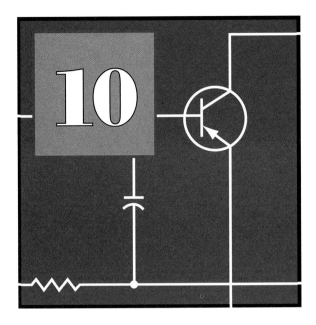

Magnetism

Objectives
After studying this chapter, you will be able to answer these questions:
1. How can an electric current produce magnetism?
2. What are the common applications of the electromagnet?

Important Words and Terms
The following words and terms are key concepts in this chapter. Look for them as you read this chapter.

ampere-turn
electromagnet
induction
left hand rule for a coil
lodestone
magnet
magnetic field
magnetic flux
magnetic material
magnetic shielding
north pole
permeability
relay
reluctance
residual magnetism
solenoid
south pole

What is the great and mysterious force called magnetism? Scientists have proposed theories upon the nature of magnetism and have learned how to use this mysterious force, but no one has ever *seen* it. You probably have a horseshoe magnet at home and have discovered that it will pick up nails and certain other kinds of metal. If you are interested in camping or hiking, you have used the compass and learned that it points in the general direction of north. The compass is one of our more common applications of magnetism.

Hundreds of years ago, the Greeks discovered small pieces of iron ore, called magnetite. They learned that these small rocks attracted small pieces of iron. Because these small stones were found near Magnesia in Asia Minor (present day Turkey), they were named *magnetite*. The word *magnetism* is derived from magnetite.

Lodestones
Centuries ago, Chinese sailors used these small magnetite stones fastened to sticks or wood floating in a container of

liquid. The stick or floating stone would turn in the direction of north, and by this means the navigator could chart the direction the ship was sailing. This was the earliest form of a compass. These stones became known as "leading stones" or *lodestones.*

A material or metal that has the property of attracting metals such as iron and steel is known as a *magnet.* The materials that it attracts are called *magnetic materials.* All magnets have two poles. The pole that is attracted toward north is called the *north pole* while the opposite pole is called the *south pole.*

Between the north and south poles exists what the scientist calls a *magnetic field* consisting of many lines of magnetic force. This magnetic field is frequently referred to as the *magnetic flux,* **Figure 10-1.**

Permanent Magnets

Permanent-type magnets are made in many shapes and sizes, Figure 10-1. They all have north and south poles and they all have magnetic fields. Materials used in the manufacture of such magnets have the ability to retain their magnetism over long periods of time. High carbon steel and special alloys, such as alnico (a mixture of aluminum, nickel, and copper), exhibit this property. Low carbon steel and iron do not retain magnetism, but they do conduct or concentrate a magnetic field very well. They have a high permeability. *Permeability* is the characteristic of a material to conduct magnetic lines of force.

You will be interested in understanding how a magnet is made. The molecular theory of magnetism explains this phenomenon by stating that in an unmagnetized steel bar, the molecules are arranged in random order as in **Figure 10-2.** When this steel bar is rubbed with a magnet or placed in a strong magnetic field, all of the molecules line up in the same direction. Each molecule assumes a north and a south pole, as shown in **Figure 10-3.**

When the molecules are in this orderly formation, the steel is magnetized. This theory is proved to some extent by the fact that if the magnetized steel bar were broken into smaller pieces, each piece of steel would be a magnet of its own.

You can try this experiment. Take a steel bar about two or three feet long. Hold it in a north-south direction and hit it several times with a hammer. The bar will become weakly magnetized and will attract a compass needle. Now turn the bar in an east-west direction and again hit it several times with the hammer. A test with your compass will now show that it is demagnetized. This phenomena can be explained when you realize that the earth itself is one

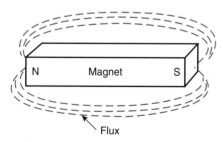

Figure 10-1.
A bar magnet showing magnetic flux.

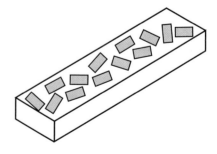

Figure 10-2.
In a nonmagnetized piece of steel, the molecules are in random positions.

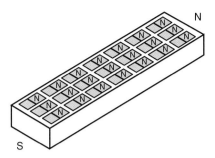

Figure 10-3.
The molecules are in line in a magnetized steel bar.

enormous magnet, and many invisible lines of magnetic force exist between the north and south pole, **Figure 10-4.**

When you turned your bar in a north-south direction, the earth's magnetic lines of force flowed through your bar because the bar has better permeability than the air about it. The bar conducts magnetism better than air. When you hit the bar, the molecules were physically jarred, and they lined up in their magnetized positions.

Permeability, the property of a material to conduct magnetism, can be compared to conductivity (G) in an electrical circuit. The resistance of a material to conduct magnetism is called *reluctance.* It can be compared to resistance (R) in an electrical circuit.

Laws of Magnetism

As you learned in Chapter 1, the electron and the proton in an atom are attracted to each other because they had opposite electrical charges. In the same way, the north pole of one magnet is attracted to the south pole of another magnet, and vice versa, because they have opposing magnetic forces. Magnetic poles that are *like* (have the same force) are repelled from each other. **Figure 10-5** is a sketch that shows how the poles react to each other.

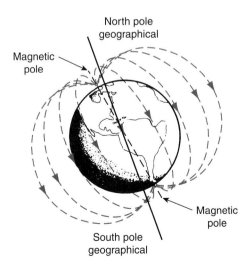

Figure 10-4.
The earth is a big magnet. A compass points toward the magnetic pole.

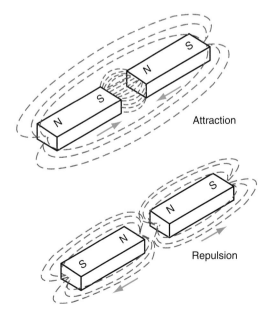

Figure 10-5.
Unlike poles attract. Like poles repel.

History Hit!

Hans Christian Oersted (1777–1851)

A Danish physicist, Oersted was the first to discover that electrical current did indeed produce a magnetic field. He studied at the University of Copenhagen and was later appointed professor of physics there. His efforts related to electricity and magnetism led a great many others into research on this topic.

Magnetic Field Pictures

It is hard to believe something that you cannot see. Let's take an ordinary bar magnet. Over this we will place a flat sheet of paper on which we have sprinkled some iron filings. Iron filings are nothing more than tiny little slivers of iron that have been filed off a larger piece of iron. By tapping the paper gently, the particles of iron will line themselves up in patterns conforming to the magnetic lines of force similar to **Figure 10-6**.

Electricity and Magnetism

For many years scientists believed there was a definite relationship between electricity and magnetism. The experiments of Hans Christian Oersted in 1820 proved that a magnetic field was produced around a conductor when a current was flowing.

You can prove this by performing the simple experiment shown in **Figure 10-7** just as Oersted did. Run a wire (insulated) through a piece of paper. Place two or more small compasses on the paper close to the wire conductor. As the current flows, the magnetic field encircling the conductor causes the compass needles to line up in the direction of the magnetic field. If you reverse the current by changing the battery terminals, the compass needles will indicate that the direction of the magnetic field has also reversed. To determine the direction of the magnetic field you can use the *left hand rule for a coil.* Grasp the conductor with your left hand. Let your thumb point in the direction of the current. Your fingers around the conductor will point in the direction of the magnetic field around the wire.

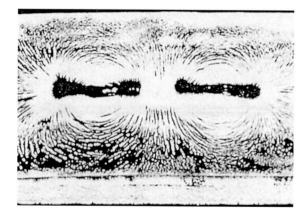

Figure 10-6.
Magnetic lines of force are shown through iron filings.

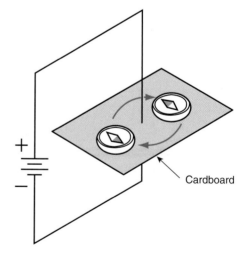

Figure 10-7.
Current flowing in a conductor creates a magnetic field around the conductor.

Were you to wind a copper conductor into a coil of several turns (if using non-insulated wire, be sure each loop does not touch), you would find that the fields around the wire combine to form a magnet.

Such a coil can be used as a solenoid. A *solenoid* uses a magnetic field to cause physical movement. The actual strength of the magnetism produced by such a coil depends upon the number of turns of wire and the strength of the current flowing through the coil. The product of the number of turns of wire times the amperes is called the *ampere-turns* of the coil. Ampere-turns is a measurement of the magnetic field strength of the coil.

Electromagnets

To improve the magnetic field strength of the solenoid, an iron core can be inserted. As iron provides a better path (higher permeability) for the lines of force than air, the strength of the magnetism is much greater, **Figure 10-8.** Such a device is known as an *electromagnet* and is used extensively in electrical/electronic equipment. You can use electromagnets in the construction of relays, doorbells, buzzers, and circuit breakers.

If the core, upon which the coil is wound, is made of iron or low carbon steel, the magnetic effect almost disappears when the current ceases to flow. Such magnetism remaining in the core is termed *residual magnetism.* In order to use the coil as a controlled electromagnet, you would wind it on a core that retains little of its magnetism.

Another principle we should consider is the strength of the electromagnet. It depends, of course, upon the ampere-turns and the permeability of the core. Electromagnets can be made to operate devices requiring a tiny force, or one sufficient to move tons of weight.

Relays

There are hundreds of applications of electromagnets in relays. A *relay* is a magnetic switch, and is usually used when it is necessary to control a rather large electric current with a small control current. See **Figure 10-9.** If you construct *Project 5—Magnetic Relay,* found in Chapter 19, the principles of controlling circuits by electromagnetism will be better understood.

Reviewing Figure 10-9, a control circuit is used in the relay coil. The coil is fixed on an iron core to enhance the magnetic field. The control circuit normally has a low voltage (12–24 volts), low current (approximately 100–500 mA) power source connected to it. When current is permitted to flow in the control circuit, a magnetic field is established. This magnetic field then causes the contacts A and B to close. Usually, one contact is fixed in place (B in this example), while the other contact (A) is on a movable armature. The armature moves in relation to the magnetic field. If there is a magnetic field present, the armature moves to the coil and closes the contacts, A and B. This completes a high-voltage, high-current circuit. Once the magnetic field is removed, a tiny spring moves the armature away from the coil. Thus, a low-voltage, low-current circuit is able to control a much higher voltage and current circuit.

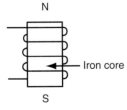

Figure 10-8.
An iron core increases the strength of the electromagnet.

82 Electricity

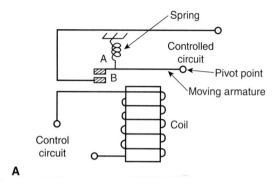

Figure 10-9.
A—A schematic diagram of a relay. B—Example of a relay.

The device is so arranged that the moving armature that has contact point B on it must be manually reset before the normal operation of the circuit can be resumed.

The Reed Relay

The reed relay is another similar application of magnetism. Referring to

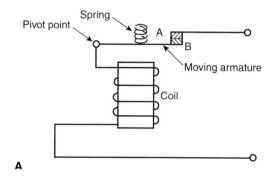

Figure 10-10.
A—A schematic diagram of the magnetic type of circuit breaker. B—A circuit breaker.

Another similar application is drawn in **Figure 10-10**. Here the coil is in series with the contact points. This is one type of circuit overcurrent protective device. As long as the current flowing in the circuit is within safe limits, the magnetic force of the coil will not overcome the tension of the spring. If the current exceeds a specified value, contact B is pulled down by its moving armature. This opens the circuit, because contacts A and B are now open.

Figure 10-11, notice that two magnetically sensitive switch contacts are enclosed in a glass tube. If a permanent magnet is brought close to the glass tube, it causes the switch contacts to close.

The reed relay can also be operated by an electromagnet. The operating coil is placed around the reed relay. Once energized, the coil's magnetic field closes the circuit by moving the reeds of the switch.

Solenoids as Switches

Solenoids, discussed earlier, can also be used to convert electricity to magnetism to mechanical motion, **Figure 10-12.**

We have already learned that when a current flows through a coil, the coil becomes a magnet and has a north and south pole similar to the permanent bar magnet. This coil attracts particles of iron in much the same manner as a permanent magnet.

If a core of soft iron is placed in the center of such a solenoid coil, it is attracted by the coil, Figure 10-12, and drawn into

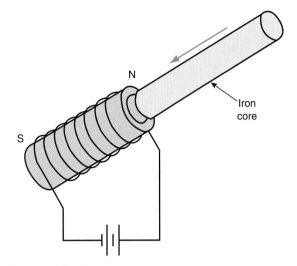

Figure 10-12.
The iron core is drawn into the coil.

the center of the coil. The solenoid will actually pull the iron core into the coil. The core's movement stops when the attracting forces of each end of the coil are balanced, **Figure 10-13,** and the core has centered itself in the coil, where the magnetic fields are the strongest.

This type of device has many applications. The movement of the iron core can be mechanically linked to switches, levers, or

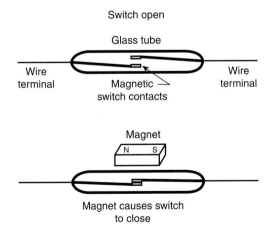

Figure 10-11.
A sketch of a reed relay. The magnet causes the reed contacts to close.

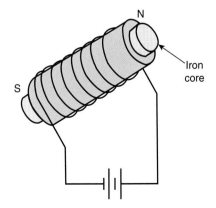

Figure 10-13.
The movement of the core stops when the magnetic forces are balanced.

gears to produce the desired mechanical action. One common application is used in the automobile, where the solenoid is used to mesh the gear on the starting motor with the gear on the flywheel of the engine. At the same time, the solenoid action closes a switch that supplies current to the starting motor, **Figure 10-14.**

Project 4—Door Chime and *Project 8—Electric Engine* in Chapter 19 are based on the principles of the solenoid. You will find them interesting and practical applications of magnetism.

Hall Effect Devices

Hall effect sensors are devices that produce output voltages in the presence of a magnetic field. In most Hall effect devices, a strip of semiconductor material is exposed to a magnetic field. A voltage is applied to opposite edges of the semiconductor. When the semiconductor material is then exposed to a magnetic field, its output voltage (taken from the other two edges of the semiconductor) is in proportion to the magnetic field. These sensors are used in various applications such as measuring speed of conveyor lines or counting.

Magnetism by Induction

Why will a magnet pick up a nail? Experimentation has disclosed that if a piece of metal, such as a nail, is brought close to a magnet, it also becomes a magnet. It assumes a polarity that will cause attraction between the nail and the magnet, see **Figure 10-15.** If a second nail is brought close to the end of the first nail, it will also be attracted, as in Figure 10-14B. This is magnetism by *induction,* which is the basis of the theory of magnetism, Figure 10-15.

Magnetic Shielding

Closely related to the phenomenon of magnetism by induction is the problem of keeping a magnetic field away from delicate testing instruments and meters. There is no known shield against magnetism. It passes through nonmagnetic materials as if they were not there. Try holding a piece of

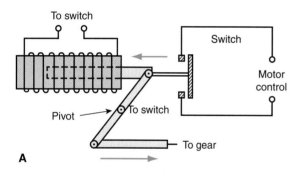
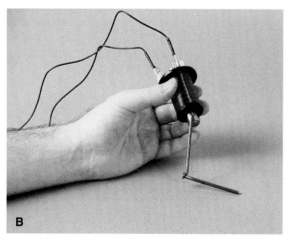

Figure 10-14.
A—The solenoid is used to close the motor switch and to engage the gears in the automotive starter. B—Magnetic induction.

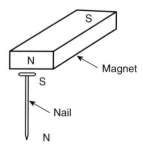

Figure 10-15.
The nail becomes a magnet by induction.

Chapter 10   Magnetism

glass between a magnet and a nail. The nail is picked up as if the glass were not there.

However, magnetism can be led around or directed away from instruments or devices needing protection using *magnetic shielding.* By placing a magnetic material such as a piece of iron in the field of a magnet, the flux lines of magnetism follow the direction of the iron, because the iron has a greater permeability than the surrounding air. The magnetic field has actually become a little stronger by this increased permeability. The field has been made to follow the iron and thus contain it while keeping it away from a less desirable area of the circuit.

Meters, transformers, and other components are frequently placed in metal cans that effectively shield them against the influence of magnetism. In radio equipment, metal fences or partitions are sometimes placed between sections of a circuit to prevent interaction between the circuits. In television equipment, shielding is also important to ensure the best operating results.

Quiz–Chapter 10

Write your answers to these questions on a separate sheet of paper. Do *not* write in this book.

1. A substance that has the ability to attract certain kinds of ferrous (containing iron) metals is known as a _____.
2. The magnetic field, consisting of many lines of magnetic force, running from the north pole to the south pole of a magnet is called the magnetic _____.
3. Draw two sketches showing the position of the molecules in an unmagnetized and a magnetized piece of steel.
4. If a magnetized bar of steel were broken into three pieces, what effect would it have on the magnetism of pieces of the steel bar?
5. The property of a substance to conduct a magnetic field is called _____.
6. The resistance a material has to conducting a magnetic field is called _____.
7. The north pole _____ the south pole.
8. Like poles _____ each other.
9. The strength of the magnetic force of a coil depends upon what two factors?
10. Why will a solenoid coil increase its magnetic strength when an iron core is inserted?
11. Why will a magnet pick up a nail?
12. What is the purpose of magnetic shielding?
13. A coil of 500 turns has a current of 2 amperes flowing through it. What is its field strength?
14. Draw a sketch of two bar magnets with the N pole of one near the S pole of the other. Draw lines showing the magnetic fields.
15. Draw another sketch of two bar magnets with the N pole of one near the N pole of the other. Draw lines showing magnetic fields.

Typical computer hard drive. Information on your computer's hard drive is stored magnetically.

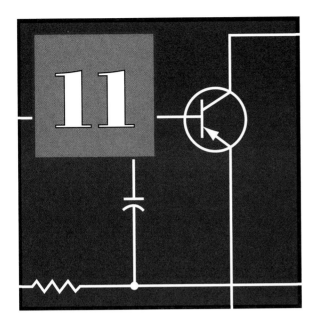

Motors

11

Objectives
After studying this chapter, you will be able to answer these questions:
1. How can electrical energy be converted into mechanical energy?
2. What is the theory and operation of the electric motor?

Important Words and Terms
The following words and terms are key concepts in this chapter. Look for them as you read this chapter.

armature	field magnet
brush	field winding
commutator	motor action
compound motor	series motor
electric motor	shunt motor

In your readings on magnetism, you learned that unlike poles of magnets are attracted to each other, but like poles repel each other. In the case of relays and solenoids, you observed that this attractive force of magnetism could be converted into mechanical action.

How the Motor Works
The *electric motor* is the most common method of converting electrical energy into rotating motion. Motors are used in thousands of ways in home and industry. They are used to heat and cool the home, to keep food cold, to wash clothes, mix food, and countless other uses. They are used in industry to provide power for cutting, shaping, and finishing materials into useful products. Our life would be a hard life and our standards of living would be much different, if the electric motor were not yet created by science and invention.

All motors operate on the principle of attraction and repulsion of magnetic poles. To understand this, we will build a simple direct current motor and trace each step of its construction to discover the method of producing rotation by the interaction of magnetic fields.

First, we must create a magnetic field, **Figure 11-1.** The familiar horseshoe magnet

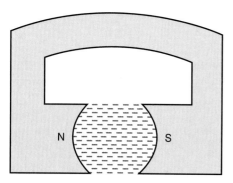

Figure 11-1.
A horseshoe magnet and magnetic field.

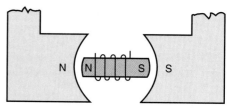

Figure 11-3.
The current flowing in the armature produces the indicated polarity.

produces the field, with lines of magnetic force running from north to south. This is known as a *field magnet.*

In our study of electromagnets, we learned that if a coil is wound on an iron core and a current passed through it, a magnet is produced. The polarity of the electromagnet depends upon the direction of the current in the winding. We will wind such a coil and put it on a round shaft about which it can rotate, **Figure 11-2**. In this way, the core can move or rotate. The core must never touch the coil, or friction (from this contact) would slow it down or stop it. Additionally, this would damage the coil insulation and can even break the wire causing an open circuit.

Assume that the current flowing creates the polarity (north, south) as indicated in Figure 11-2. This assembly is called the *armature.*

In **Figure 11-3**, the field magnets and the armature have been combined into one assembly. Notice the polarity of both the field (stationary or nonmoving) magnet and armature (moving or rotating magnet).

Now a north pole repels a north pole and a south pole repels a south pole, so the armature turns one-quarter turn (90 degrees of rotation) to the position shown in **Figure 11-4**. The arrows indicate the direction of rotation.

Since north and south poles attract each other (unlike poles attract), the armature turns another quarter turn (another 90 degrees of rotation) and comes to rest in the position shown in **Figure 11-5**. No further rotation will take place. The armature is locked magnetically in place. Remember that the polarity of the armature depends upon the direction of the current.

What would happen if, just as the north and south poles of the field and

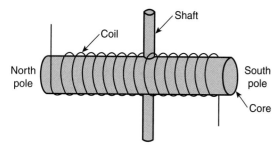

Figure 11-2.
An electromagnet is wound and placed on a shaft so that it will revolve.

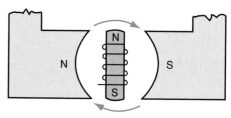

Figure 11-4.
North repels north and south repels south.

Chapter 11 Motors

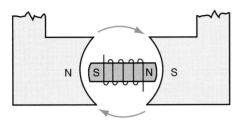

Figure 11-5.
South attracts north and north attracts south.

armature magnets are lined up (Figure 11-5), the current in the armature reversed itself? This can be accomplished several ways, using a switch or using a motor controller device. If this current reverse were to happen, the situation would suddenly appear as in **Figure 11-6.**

This is the same position as when we started in Figure 11-3. Like poles repel each other for a quarter turn and then the unlike poles attract each other for the final quarter turn. One complete revolution of the armature (360 degrees of rotation) has been accomplished. By reversing the current in the armature again, the rotation continues. As long as the current is reversed each one-half revolution, the rotation will continue indefinitely. **Figure 11-7** shows graphically the chain of events just described. Note that motor action of this type is based upon magnetic attraction and repulsion. Remember, for rotation to continue, the current in the armature must be reversed.

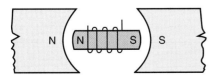

Figure 11-6.
The armature has rotated one revolution. At this point, the current in the armature is reversed. The armature will continue to rotate.

History Hit!

Nikola Tesla (1856–1943)

Tesla was born in Croatia and studied electrical engineering. Later, he emigrated to the United States. For a time, Tesla was employed by Thomas A. Edison. However, disagreements arose over Tesla the theorist and Edison the practicalist. It was through partnership with George Westinghouse that Tesla's development in direct current (dc) motors, alternating current (ac) motors, and power transmission and generation became well known. He spent his later years experimenting with high frequency ac. Tesla's contributions to the motors that power industry are well known. Later in life, he became a bit of an eccentric and recluse. Yet, his design of ac power transmission is still in use today!

The Commutator

It is necessary to change the direction of the current in the armature at just the correct point to produce the continuing rotation. Of course we cannot do this by hand with a switch. Some mechanical means is necessary to do so. The device used to change the current in the armature is called a *commutator*.

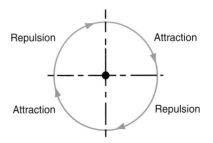

Figure 11-7.
This circle describes the chain of events during one complete revolution of the armature.

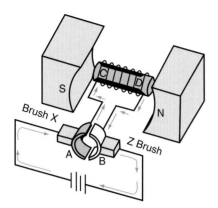

Figure 11-8.
The indicated current makes end C of the armature a south pole.

In **Figure 11-8**, the ends of the armature coil are brought out to two pieces of metal, which are insulated from each other. Notice the two pieces of metal, which form a structure called a commutator, are like a circle that has been split in half. Sliding on these metal sections are two fixed contacts known as **brushes.** A battery is connected to the brushes. Arrows indicate the direction of the current.

An electrical brush is made up of carbon that has tiny pieces of copper mixed inside of it. The carbon and copper are heated and formed into a square or rectangular shape that usually has a spring extending from it. It is this brush assembly that forms the electrical connection from the moving metal piece (commutator) to the fixed or nonmoving electrical circuit.

Current flows from the battery to the brush (X), to commutator bar A, into the armature coil and out to commutator bar B, then through the brush (Z), and then back to the battery.

Such current makes end C of the armature a south pole and D a north pole. The repulsion and attractive force causes the armature to rotate one-half turn (180 degrees of rotation). The commutator bars then turn with the armature. The position of the armature after the half turn appears as in **Figure 11-9**. Now the current flows from the battery to brush X to commutator bar B, into the armature coil and out through commutator bar A, through the brush Z, and back to the battery. Such a current will make end C of the armature a north pole and D a south pole. In other words, the commutator has changed the direction of the current through the armature. This occurs each half revolution and the conditions for continued rotation have been satisfied.

Remember, like magnetic poles repel, unlike poles attract. With the addition of the commutator (sometimes referred to as a split ring), the switching of electrical current is achieved. This switching action keeps the motor turning. The brushes are used to complete the electrical circuit between the moving commutator and the stationary electrical connections.

Motor Circuits

In the explanations that follow, the diagram shown in **Figure 11-10** is used to designate a motor. The various parts are named.

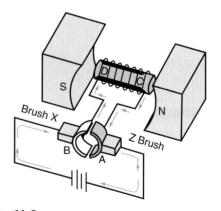

Figure 11-9.
After one-half revolution, end D becomes a south pole.

Chapter 11   Motors

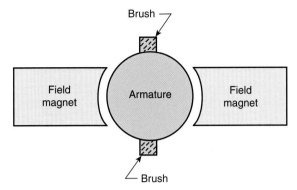

Figure 11-10.
A diagram showing the major parts of a motor.

To improve the rotating force of a motor, wind coils around the field magnets and make them strong electromagnets, known as *field windings.* A *series motor* appears in **Figure 11-11.** In the series motor, all of the current flows through the armature windings and the field windings. Trace the wiring from the negative terminal to the positive terminal. There is only one path for current to flow. Such a motor varies its speed according to the load it must turn. It has the ability to start under heavy loads. Due to their heavy starting load, automobiles use series motors as starter motors. This heavy load is found when starting the engine.

A second type of dc motor is the parallel wired motor, also called the *shunt motor,* **Figure 11-12.** In the shunt motor, the current divides. Part of the current flows through the armature windings and the remainder through the field windings. Such a motor tends to keep a constant speed under varying load conditions. It is very well suited to the operation of industrial automation devices and various machine applications.

There is another type of motor using both series and shunt windings that is known as the *compound motor.* Detailed information about all of these dc motors can be found in textbooks on advanced electricity and its applications in industry.

Further improvements are made upon our simple motor for commercial use. In studying the sequence of events as a motor completes one revolution, you will readily see that the turning force created by the interacting fields could be greatly improved if more field and more armature poles were used. The repulsive and attractive forces of magnetic fields decrease rapidly as the distance between them is increased. Mathematically speaking, the magnetic

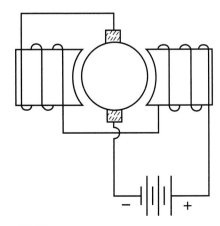

Figure 11-11.
The diagram of a series wound motor.

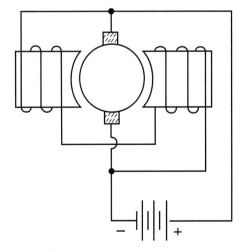

Figure 11-12.
The diagram of a shunt, or parallel, wound motor.

force between two poles varies inversely as the square of the distance between them. So by using more poles in a motor, the distance between poles is decreased and as a result the turning force, or torque, is increased. This is a very important concept related to electric motors.

The Complete Motor

In a complete motor, the field windings are wound on the field poles, and the armature and commutator are mounted on a shaft. The whole assembly is placed in a shell or case. Suitable bearings are provided for the armature shaft. The case is provided with a means of holding the brushes under spring tension. This is so that they will constantly ride on the commutator bars. As the bars and brushes wear, the springs keep a constant brush pressure and ensure a good electrical contact.

Many motors have built-in fans that circulate air through the motor keeping the motor cool during operation. This is just one of many features that we have not completely discussed. Motors can be complicated and require advanced knowledge to properly apply them in industry.

Project 7—Two-Pole Motor Experiments (in Chapter 19) describes in detail the construction of an experimental two-pole motor. This motor can be connected either as a series or a shunt wound motor. Study the action of the motor. Observe the function of the commutator.

Motor Principles

The simple two-pole motor described in this text and constructed in *Project 7* is frequently called the St. Louis motor. Although this motor depends upon the interaction between the magnetic fields of the armature and the field windings, a more accurate description of the modern motor must include the action of a current carrying conductor in a magnetic field.

Reviewing Chapter 10 on Magnetism, you will recall that a magnetic field exists around a conductor carrying an electric current. The left hand rule states that if you grasp the wire by your left hand with your thumb pointing in the direction of the current, your fingers curl around the wire in the direction of the magnetic field. Symbols currently used to represent this phenomenon are shown in **Figure 11-13**. The dot in the center of the circle on the left indicates the point of an arrow. The current is flowing toward you. The cross in the circle on the right represents the feathers on the arrow as it goes away from you or into the paper. The circular arrows indicate the direction of the magnetic field in each instance.

When such a current carrying conductor is in another magnetic field, there is interaction between the fields that will cause motion. Observe this action in **Figure 11-14**. The arrows show the direction of both the fixed or stationary permanent magnet field and the field around the moving conductor. On the upper side, the conductor field opposes the permanent field (arrows in opposite directions). On the lower side, the conductor field reinforces the permanent field (arrows in same direction). The conductor moves toward the weakened field, or upward.

Figure 11-13.
Symbols used to show current and magnetic field.

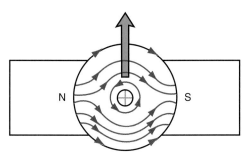

Figure 11-14.
The interaction between a magnetic field and a current carrying conductor.

Coils of wire or conductors are used in the armature of a motor. The *motor action* between the two fields causes the rotation. In the dc motor, the current must be periodically reversed by means of a commutator to provide continuous rotation. A review of **Figure 11-15** shows the many parts of a dc motor that we have discussed in this chapter.

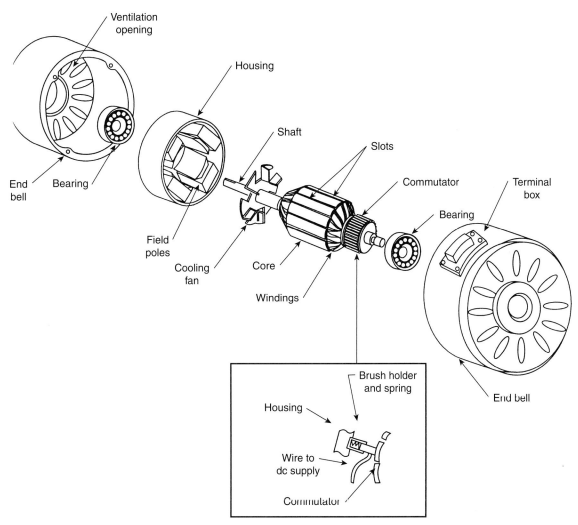

Figure 11-15.
Parts of a dc motor.

Quiz–Chapter 11

Write your answers to these questions on a separate sheet of paper. Do *not* write in this book.

1. The rotating coils of the motor are called the _____.
2. A device to reverse the current in the armature windings is the _____.
3. Draw a diagram of a series motor.
4. List two characteristics of the series motor.
5. Draw a diagram of a shunt or parallel wired motor.
6. List two characteristics of the shunt motor.
7. Fixed contacts that are used to make electrical connections to the moving armature are called _____.
8. List twelve devices or pieces of equipment that use electric motors.

Interior of a stepper motor. This motor rotates in very small increments. They are commonly used in disk drives and robots.

Direct Current Generators

Objectives

After studying this chapter, you will be able to answer these questions:
1. How is mechanical energy converted to electrical energy?
2. What is the theory and operation of the direct current generator?

Important Words and Terms

The following words and terms are key concepts in this chapter. Look for them as you read this chapter.

alternating current (ac)
compound generator
copper losses
direct current (dc)
eddy current losses
galvanometer
generator
hysteresis loss
independently excited
laminating
left hand rule for a conductor
pulsating direct current
series generator
shunt generator

A generator can be considered as the opposite of the motor. A motor converts electrical energy to mechanical energy; a *generator* converts mechanical energy to electrical energy. The actual construction of both devices is very similar.

For many years, scientists were familiar with the electromagnet. They knew that electricity could produce magnetism. It was not until the experiments of the renowned Michael Faraday in 1831 that the means was discovered to change magnetism into electricity.

In order to produce an electric current from magnetism, three conditions must exist.
1. A magnetic field must be present.
2. A conductor (wire or coil of wire) must be available.
3. There must be relative motion between the field and the conductor. Either the field must move or the conductor must move.

You can perform a simple experiment to prove this principle. Wind several turns of wire into a coil about two inches in diameter. Connect the ends of the coil to an analog

History Hit!

Michael Faraday (1791–1867)

Faraday was a British chemist and physicist. He began his career as a bookbinder, and while at his craft he was able to read many books related to chemistry and science. A chance meeting with Sir Humphry Davy of the Royal Institution put Faraday in touch with some of the greatest scientific minds of the era. Beginning his activities in chemistry, he later ventured into electricity. When finished he would invent the electric motor, dynamo (a dc generator made with a commutator to convert mechanical energy into electrical energy), and the transformer. The unit measure of capacitance, the farad, is named for him.

the conductor and the magnetic flux itself. In order to determine the direction of the current, the *left hand rule for a conductor* is used. First, hold the thumb, index finger, and middle finger of your left hand at right angles to each other. The thumb points in the direction of movement of the conductor. The index finger points the direction of the magnetic flux (from the north pole to the south pole). The middle finger points the direction of the current. Referring to **Figure 12-1,** when the conductor is moving downward (thumb down), index finger pointing to the south pole (to the right), the middle finger points away from you. Electrical current is flowing away from you. If the conductor is moving upward (thumb up), index finger pointing to the south pole (to the right), the middle

galvanometer. A *galvanometer* is a very sensitive meter that can indicate small values of current or voltage. The galvanometer also has a zero center point. This means that instead of zero being found at the left hand side of the scale as it is on many types of analog meters, zero is found at mid-scale. This way, the meter can provide for a negative or positive indication.

Now pass the coil back and forth between the poles of a horseshoe magnet. Notice how the indicating needle moves. As you pass the coil down through the magnetic field, the indicator moves in one direction. As you bring the coil up through the magnetic field, the indicator moves in the opposite direction.

Direction of Current

It is frequently necessary to determine the direction of the current in a conductor as it passes through a magnetic field. This depends on the direction or movement of

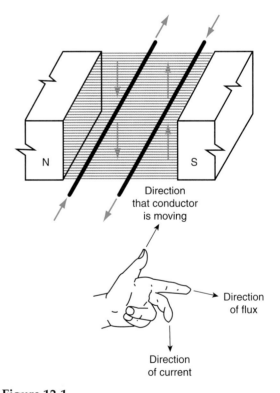

Figure 12-1.
The direction of current as the conductor passes through a magnetic field.

Chapter 12   Direct Current Generators

finger points toward you. In this case, electrical current is flowing toward you.

One other consideration is the intensity of the current. It is easy to see that no current flows when the conductor is outside of the magnetic field. As the conductor first enters the field a little current starts to flow. The current increases as the conductor moves toward the halfway point or to the center, where the magnetic field is the strongest. Then, as the conductor moves away from the center, the current decreases until it becomes zero when the conductor has passed out of the field on the other side. **Figure 12-2** shows this current graphically.

To further develop our graph we can add the curve that shows the current as the conductor is moved down through the field. When drawing such a graph, the curve above the zero line indicates current in one direction (positive) and the curve below the zero line (negative) shows current in the opposite direction. **Figure 12-3** shows the complete graph of current made by one upward movement and one downward movement of the conductor through the magnetic field.

The Generator

In order to create a continuous motion of a conductor through a field, it is convenient to have a coil rotate in the field. In

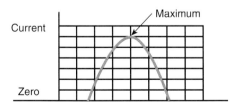

Figure 12-2.
Current starts at zero, rises to maximum amperage, and returns to zero as the conductor cuts across the magnetic field.

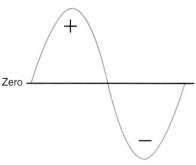

Figure 12-3.
A graph showing the current for one upward and one downward movement of the conductor.

Figure 12-4, a single loop of wire is arranged to rotate in the magnetic field. The curved arrow shows the direction of rotation of the loop of wire. As side A of the coil passes up through the field, a current flows in the direction indicated by the arrows (use your left hand rule to determine direction). At the same time, side B is moving down through the field. Current flows into side B and out of side A. As the coil turns one-quarter turn, the sides of the

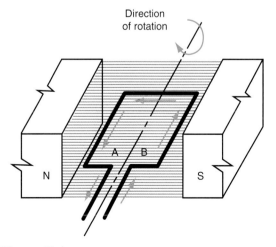

Figure 12-4.
A rotating loop is placed in the magnetic field.

98 Electricity

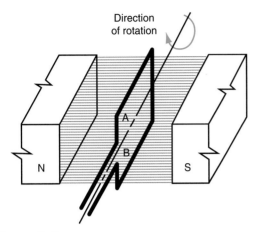

Figure 12-5.
No current is induced when the coil moves parallel to the magnetic field. The coil, or loop of wire, is no longer in the magnetic field.

loop coil are parallel to the lines of the field and are not cutting through them. At this point the current is zero, **Figure 12-5.**

As the coil continues its rotation, side A cuts down through the field and side B cuts up through the field. A current flows, but in the opposite direction. It flows out of side B and into side A, **Figure 12-6.**

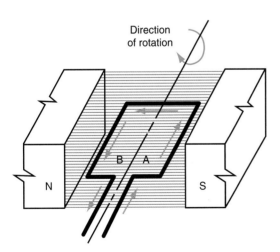

Figure 12-6.
Maximum current is induced when the coil cuts across or moves perpendicular to the magentic field. Here the coil, or loop of wire, is cutting through the maximum amount of magnetic lines.

Continuous rotation produces a current in the coil (armature), which is reversing its direction every half revolution of the coil (180 degrees of rotation). A current that changes its direction periodically is known as an *alternating current* or *ac.* Most generators are alternating current generators. However, some have an output of *direct current* (dc).

The Commutator

You will notice a similarity between the dc motor commutator and the dc generator commutator. They both serve the same purpose. Figures 12-4 and 12-6 have been redrawn in **Figures 12-7** and **12-8** to include the commutator split ring, the brushes, and an external circuit.

To understand the commutator action, trace the current shown by the arrows. In Figure 12-7, side A of the loop is cutting up through the field and side B is cutting down through the field. Current flows out

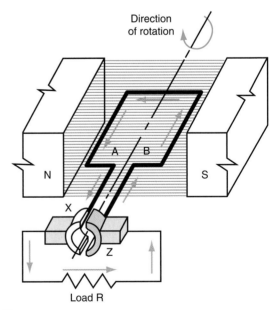

Figure 12-7.
Current flows into side B and out of side A of the coil.

Chapter 12   Direct Current Generators

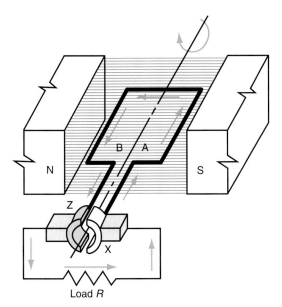

Figure 12-8.
Current flows into side A and out of side B of the coil.

of side A to commutator bar X. The brush collects current and it flows through the external circuit (load resistor *R*), to commutator bar Z, and into side B of the coil. Remember, the brushes serve as a fixed electrical connection to the moving comutator.

As the coil and commutator sections rotate through one-half turn, the situation is like Figure 12-8. The current is reversed in the loop coil, but the commutator bars have also changed position. Now the current flows out of B to bar Z, through the external circuit, to bar X, and into side A of the coil. Follow the current arrows to determine electrical current in the circuit.

It is important to remember, that although the current is alternating in the coil, the current in the external circuit is flowing in only one direction. This is referred to as *direct current* or *dc*. Using a graph to show the current, it appears as pulses of current flowing all above the zero line, **Figure 12-9**. This is called *pulsating direct current*. Note that the current stays above the zero line, it may vary in level or amplitude, but it continues to flow in only one direction.

During this discussion, we have constructed a simple dc generator. Two improvements must be made to make it more practical and effective.
1. A stronger current must be generated.
2. A smoother current (less variation) must be generated. A pulsing direct current is not satisfactory for many kinds of equipment.

In other words, we must find a way to prevent current from dipping back down to the zero line. At the zero line, there is no electrical current flowing.

Generator Output

The simple generator that has been described uses only a single loop of wire rotating in the magnetic field. The currents generated are weak. This is because of current levels falling to zero three times in one complete rotation of the wire loop. To improve the output, the rotating loop of wire or coil can be made of many turns of wire. Such is the case in the armature of a practical generator. Likewise, the magnetic field in which the coil or armature rotates can be made a great deal stronger with the use of electromagnets instead of permanent magnets. These electromagnets are made by

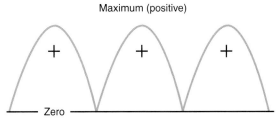

Figure 12-9.
The waveform of the current in the external circuit of the direct current generator.

wrapping copper wire on iron cores that are fixed in position by the coil or armature. These field windings create a strong magnetic field once energized.

The examination of Figure 12-9 discloses that a generator using a single rotating coil for an armature generates two pulses of direct current per revolution of the coil or armature. By increasing the number of coils in the armature, with a corresponding increase in commutator bars, a greater number of direct current pulses can be obtained from one revolution of the armature.

In **Figure 12-10,** the curves show the current from a two-coil armature. Notice that as the current starts to drop in the first coil, the current is rising in the second coil. The resulting current has a lesser degree of variation and it is easier to make into a pure direct current, **Figure 12-11.**

Commercial generators have many windings on the armatures and produce almost a pure direct current.

Generator Losses

All of the current produced by the generator does not serve a useful purpose. There are losses within the generator itself. These losses usually take the form of dissipated heat.

In our study of resistance, we learned that all conductors offer some resistance to the flow of current. In a generator, many feet of copper wire are used in the armature and field windings. This resistance creates a loss of energy and takes the form of heat. It can be measured in watts of power lost:

$$P = I^2R$$

where P is the power, I^2 is the current squared, and R is the resistance.

The losses resulting from this source are known as *copper losses.* This power loss equals the power lost due to the heating of the wire used in the generator.

A second important loss in the generator is known as *hysteresis loss.* Hysteresis sometimes is defined as molecular friction due to the changing magnetic field. With the rotation of the armature, the groups of molecules in the core of the armature are constantly changing magnetic polarity. This internal changing polarity creates a heat

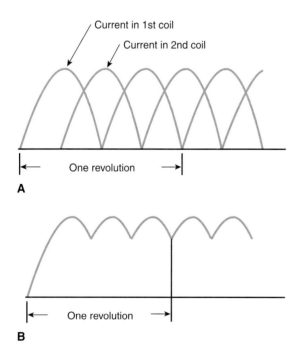

Figure 12-10.
A—The waveform in the external circuit of the generator with a two-coil armature.
B—Compare this waveform with Figure 12-9. In this waveform, current does not drop to zero, and variation is greatly reduced.

Figure 12-11.
Pure direct current (dc) as found in a battery.

within the core, which is a loss. Special alloy steels and heat-treating processes have been discovered that greatly reduce the hysteresis loss of the armature core.

A third loss in the armature is known as the *eddy current loss.* This loss can be understood when one realizes that not only is a current being generated in the windings of the armature, but also currents are being generated in the core on which the windings are wound. This is because the cores are usually made of iron or iron-type materials. Iron can conduct electricity. A current can therefore be generated in a core if the magnetic field varies. These currents circulate back and forth in the core. Eddy current losses are overcome by laminating the core of the armature. *Laminating* the core means that the core has been sliced into very thin narrow pieces, as opposed to one solid block of iron. These thin slices of metal are then insulated from one another. This prevents a sizable electrical current from flowing in the conductive iron core.

In summary, the losses described (copper, hysteresis, and eddy current), can be found in every generator. These losses can be minimized, but never totally eliminated. Remember that copper losses are due to the resistance of the wire used. Hysteresis loss comes about due to the heating of the core

> **Web Wanderings!**
>
> http://www.howstuffworks.com/
>
> Satisfy your curiosity and get the "inside" story of how electronic devices and scientific principles work. How Stuff Works covers a wide range of categories, such as Auto Stuff, Computer Stuff, Electronics Stuff, and Science Stuff.

by the variation in the magnetic field that surrounds the core. Eddy current loss is formed as an electrical current is induced (magnetically created) in the core.

Field Excitation

For the generator to operate, it must have a magnetic field. Permanent magnets can be used to create this magnetic field. But this magnetic field is weak and electromagnets have been introduced to replace them. An electromagnet requires electricity for its magnetic field to be created.

The field windings of a generator can be *independently excited* by a direct current source such as a battery. Using the same

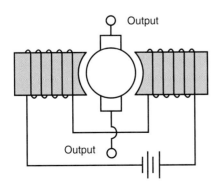

Figure 12-12.
Diagram of a generator with external field excitation (electromagnets).

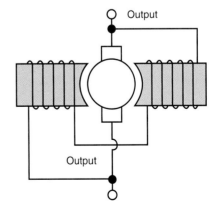

Figure 12-13.
Diagram of a shunt wound generator. The output connections are part of the electromagnetic (field) circuit.

diagrams as we used for a motor, this circuit would appear as in **Figure 12-12**.

A second method of exciting the fields that is widely used in industry is called a *shunt generator.* The generator gets its start from a small amount of residual magnetism left in the field poles (iron cores upon which the wire is wrapped). As current builds up, a part of its output is used to excite its field windings. **Figure 12-13** is an example of this circuit.

A generator can also be connected with the field in series with the armature. This creates a *series generator,* but it is not a popular method for the generation of power.

A combination of series and shunt connections does have many applications in commercial generators. This is known as a *compound generator.*

Quiz–Chapter 12

Write your answers to these questions on a separate sheet of paper. Do *not* write in this book.

1. Who invented the electric motor?
2. What is meant by copper losses?
3. What is meant by eddy current loss?
4. What is hysteresis loss?
5. How are eddy current losses minimized?
6. Classify generators into four types, depending on method of field excitation.
7. What is the purpose of the commutator in the generator?
8. In order to induce current flow in a conductor, there must be relative motion between the conductor and a _____.
9. What is the advantage of having several coils of wire on the armature?
10. Draw a circuit diagram of a shunt generator.
11. How does a shunt generator get its initial field excitation?
12. If a variable resistor were connected in series with the field windings of a generator, it would be possible to manually control the output of the generator. Explain why this is true or untrue.

Alternating Current

Objectives

After studying this chapter, you will be able to answer these questions:
1. How does an ac generator differ from the dc generator?
2. What is a sinusoidal (sine) waveform?
3. What is meant by phase relationships?
4. How can dc current be compared with ac current?

Important Words and Terms

The following words and terms are key concepts in this chapter. Look for them as you read this chapter.

average value	root-mean-square (rms) value
cycles per second	sine wave
effective value	single-phase generator
frequency	
hertz (Hz)	sinusoidal waveform
peak-to-peak value	slip ring
peak value	three-phase generator
period	
phase displacement	vector

In the chapter on direct current generators, we learned an alternating current was generated in the rotating armature. However, due to the commutator, the alternating current (ac) appeared as direct current (dc) in the output of the generator.

The most common form of electrical current used is alternating current. It is used in industry to run motors that operate machines. It is used in the home for lighting, heating, cooling, cooking, and entertainment purposes.

Many forms of electrical power plants are located throughout the United States. A power generating plant converts various forms of energy into electrical energy. A fossil fuel plant burns coal to make steam. Many power generation facilities create fossil fuel to drive turbines (a fan-like assembly used to catch moving steam), which then rotates a generator. Steam can be created by any fossil fuel: coal, oil, and natural gas.

The energy of falling water is captured in a hydroelectric plant and changed via turbines and generators into electrical energy. Nuclear power plants heat water using the heat generated by the fission of uranium fuel, **Figure 13-1.** The steam from the heated water drives a turbine that is joined to a generator and rotates the armature to create electricity.

The ac generator is similar to the simple dc generator with one exception. The commutator has been replaced with *slip rings*, **Figure 13-2.**

A slip ring is connected to each end of the coil. The brushes are in constant contact with the rotating rings providing a way to pick up the electrical energy generated in the rotating armature. Note that slip rings remain in contact with each coil connection—no switching (commutation) takes place.

The changing polarity and changing direction of current in the rotating armature has already been explained in the unit on the principles of the generator. The two important things to remember in the case of the ac generator is that:

1. The current in an external circuit connected to the brushes of the generator is an alternating current periodically rising and falling and changing direction.

Figure 13-1.
The Calvert Cliffs Nuclear Power Plant in Lusby, Maryland. (Constellation Nuclear LLC)

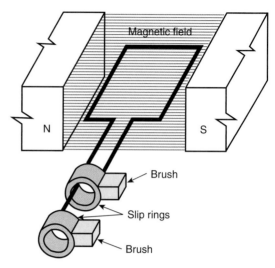

Figure 13-2.
The alternating current generator has slip rings instead of a commutator.

2. The same rising and falling and changing of current is also found in the armature.

Vectors and Sinusoidal Waveforms

Engineers and technicians have a convenient way of representing a force. It is represented by means of a vector. A *vector* is simply an arrow. The length of the arrow represents the magnitude (level or amount) of the force. The direction that arrow is pointing is the direction in which the force is acting. Sometimes this technique, used to combine variables, is called *graphical mathematics*.

The vector can be used to represent the magnitude of a generated voltage and the direction in which electromotive force is acting or causing a current to flow. To better understand the use of vectors, refer to **Figure 13-3** and the generation of the sinusoidal or sine waveform.

For example, let's assume the generator can produce a voltage of 50 volts. The vector

Chapter 13   Alternating Current

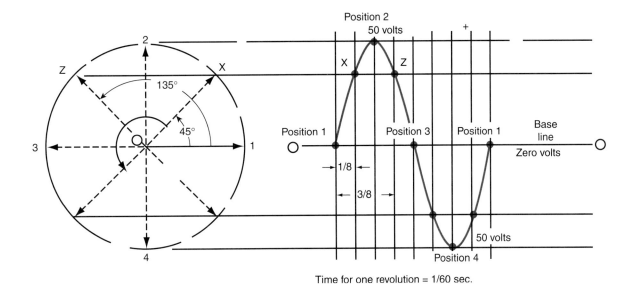

Figure 13-3.
The development of a sinusoidal, or sine, waveform.

rotates around point 0 in a counterclockwise direction. This can be compared to the rotating armature of the generator. **Figure 13-4** shows the rotating armature in four positions. Now follow the explanation using both Figures 13-3 and 13-4.

Position 1: *0 Degrees of Rotation.* Conductor not cutting through the magnetic field. No voltage generated.

Position 2: *90 Degrees of Rotation.* Conductor in center of the magnetic field. Maximum voltage. Positive polarity generated.

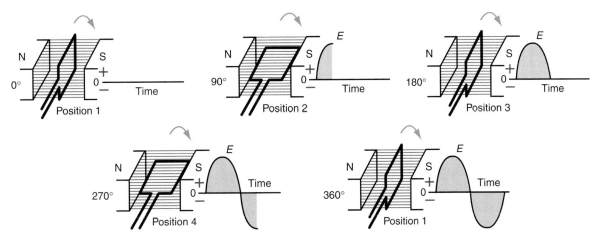

Figure 13-4.
Compare the pictures of the vector and the rotating armature coil of the ac generator.

Position 3: *180 Degrees of Rotation.*
Conductor not cutting through the magnetic field.
No voltage generated.

Position 4: *270 Degrees of Rotation.*
Conductor in center of the magnetic field.
Maximum voltage generated.
Negative polarity generated.

Position 1: *360 Degrees of Rotation (0 Degrees).* Conductor not cutting through the magnetic field.
No voltage generated.

The positive voltage is a force in one direction and the negative voltage is a force in the opposite direction. Current also corresponds to the direction of the force or voltage.

Assuming that the armature is revolving at a speed of 60 revolutions per second, the elapsed time for one revolution would be one-sixtieth (1/60) of a second. Refer to Figure 13-3 again. Note that one complete waveform occurs in 1/60 of a second. Therefore, the voltage starting at zero rises to maximum force in one direction (Position 2) then at the end of 180° rotation returns to zero, (Position 3). In the next 90° of rotation the voltage rises to maximum but in the opposite direction (Position 4). In the last 90° the voltage returns to zero (Position 1).

The waveform in Figure 13-3 is a graph of the instantaneous voltages and represents the rise and fall of the voltage during one revolution. Instantaneous values are found when voltage (or current) levels are determined at precise instances of the waveform. Points X and Z have been plotted. The waveform is the result of a line drawn through several plotted points. Point X is taken at 45° revolution and at 1/8 of the time base distance. Point 2 is taken at 90° revolution and 1/4 of the time base. Point Z at 135° revolution and 3/8 of the time base.

> **History Hit!**
>
> Heinrich Rudolph Hertz (1857–1894)
>
> Hertz was a German physicist who became a university professor. Much of his research centered along radio waves. In fact, he was the first to demonstrate the existence of such waves. Unfortunately, he died before Guglielmo Marconi turned radio wave transmission into global communication. Hertz is, however, honored for his contribution to electronic communication by having the unit of frequency named in his honor.

Many points can be plotted to get a very accurate curve. The curve is known as the *sinusoidal waveform* and represents an alternating current or voltage. We simplify the word sinusoidal by calling the waveform a *sine wave.*

The term sine wave will give you an introduction to trigonometry. It shows that the instantaneous voltage developed at any point on the sine wave depends upon the degrees of rotation of the vector.

Frequency

The chain of events described in the previous examples represents *one cycle* of operation. You will notice that from position 4 (270 degrees of rotation), the vector returns to position 1 (0 degrees of rotation) and the whole sequence of events repeats over and over for each revolution of the vector. The sine wave is a graphic picture of the instantaneous voltages during one cycle of generation. If the time duration of one cycle is 1/60 of a second, then the *frequency* of the generated voltage is 60 *cycles per second.*

The measurement of cycles per second has been replaced by the term *hertz,* abbreviated *Hz.* One hertz equals one cycle per second. Electrical power generation in the United States is based upon the 60 Hz frequency standard.

Period

The time duration of a wave form is called its period. The *period* of a wave form is the inverse of the frequency. A 60 Hz wave form has a period of 1/60 of a second. Therefore an equation can be written which tells us that:

$$\text{period (in seconds)} = \frac{1}{\text{frequency (in hertz)}}$$

Or for frequency:

$$\text{Frequency (in Hertz)} = \frac{1}{\text{Period (in seconds)}}$$

Math Manipulation!

Find the period of a 1000 Hz wave. Its period would be:

$$\text{period} = \frac{1}{1000 \text{ Hz}}$$

or 0.001 second (1 ms)

A generator with a single coil armature, as in Figure 13-4, produces electricity with only one phase, as in Figure 13-3. This is a *single-phase generator.*

It is possible to place a second coil in the generator that would produce another voltage. Usually this coil is so spaced that the second voltage is 90° out of step (out of phase) with the first voltage. The waveform output of this generator is shown in **Figure 13-5.** Voltage A is 90° out of phase with voltage B, or there is a 90° *phase displacement* (difference) or shift between the two voltages. Such a generator is a *two-phase generator.* The output circuits can be connected to use voltage A, voltage B, or both voltages in series.

A third very common type of alternator or generator is three-phase. A *three-phase generator* has coils spaced 120° apart and three separate voltages are generated. Each voltage has a phase displacement of 120°. The output wave is shown in **Figure 13-6.** Voltage B is 120° out of phase with voltage A, and voltage C is 120° out of phase with voltage B. The term phase is sometimes written with the symbol ϕ. Examples of such use are: 1ϕ equals single phase and 3ϕ equals three phase.

Average Values

A closer look at the sine wave for *one-half* cycle or one alternation discloses that the voltage is continuously varying. It

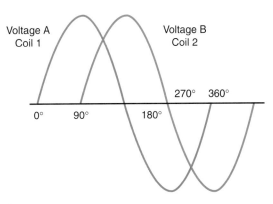

Figure 13-5.
The waveform of a two-phase alternating current. This is not a common generator type.

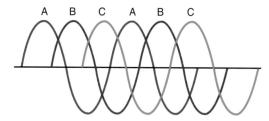

Figure 13-6.
The overlapping waveform of three-phase alternating current.

starts at zero, rises to a maximum or *peak value,* and then returns to zero, **Figure 13-7.** The value here is a voltage, but it could be a voltage, current, or power value. The overall value of this voltage waveform cannot be considered as equal to its peak voltage, because the peak is reached at only one point during the half cycle. One can find an *average value* (voltage, current, or power) represented by a sine wave by the formula:

$$E_{average} = 0.637 \times E_{peak}$$

or

$$E_{peak} = 1.57 \times E_{average}$$

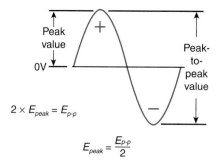

Figure 13-7.
A waveform showing the instantaneous peak and peak-to-peak voltage value.

Math Manipulation!

What is the average value of an alternating voltage that has a peak voltage of 50 volts?

$$E_{average} = 0.637 \times E_{peak}$$
$$E_{average} = 0.637 \times 50 \text{ V}$$
$$E_{average} = 31.85 \text{ V}$$

Carefully note that the average value is computed as the average value of one-half cycle. The average value of a complete cycle would be *zero.* Remember, one complete cycle has an equal amount of waveform above the zero reference line (base line) and below it. In this way, the positive alternation cancels out the negative alternation and the net result is zero.

Effective Values

A more meaningful value of an alternating current is its *effective value.* To determine the effective value, the alternating current is compared to a direct current. In other words, if a direct current will produce a certain power, what equivalent alternating current will produce the same power or work performed in an electrical circuit?

If a load, such as a resistor, were connected across an ac circuit, the instantaneous currents between zero and peak could be taken and the effective power could be found. As power equals:

$$P = I^2 R$$

all instantaneous current values could be squared and added together. Dividing the sum by the number of current values used gives the average of the squared currents. The square root of this value is known as the *root-mean-square (rms) value. Effective values* and *rms values* are the same thing. Either term can be used. Effective values are 0.707 times the peak value.

$$E_{eff} \text{ (rms)} = 0.707 \times E_{peak}$$

or

$$E_{peak} = 1.414 \times E_{eff} \text{ (rms)}$$

Math Manipulation!

Find the effective value of a 20 ampere peak current.

$$E_{eff} = 0.707 \times E_{peak}$$
$$E_{eff} = 0.707 \times 20 \text{ A}$$
$$E_{eff} = 14.14 \text{ A}$$

A 20 ampere peak alternating current has an effective value equal to 14.14 amps of direct current.

Remember the importance of the root-mean-square (rms), or effective, value of the voltage or current. These values provide the same work (heating effect) in a resistive circuit as dc. For this reason, most meters are designed to provide you rms, or effective, values for all ac measurements (voltage and current). These values are used for all home and most industrial applications of ac.

A summary of these formulas for average and effective values is as follows:

$$E_{eff} = 0.707 \times E_{peak}$$
$$E_{peak} = 1.414 \times E_{eff}$$
$$E_{avg} = 0.637 \times E_{peak}$$
$$E_{peak} = 1.57 \times E_{avg}$$
$$E_{eff} = 1.11 \times E_{avg}$$
$$E_{avg} = 0.9 \times E_{eff}$$

Peak-to-Peak Values

Another measurement value of an ac wave to recognize is *peak-to-peak value.* Study Figure 13-7 and note these values. The peak-to-peak value of a waveform always appears on an oscilloscope, which is one of the basic instruments used in observing and measuring ac waves.

The oscilloscope, generally referred to as "the 'scope," is a graphical voltmeter. Basically, the oscilloscope will display waveforms of both dc and ac over a variable time base. This means the waveform can be spread out horizontally for very fast or very slow signals. Additionally, the waveform displayed can be adjusted vertically (magnitude) for very small or very large signals. Not a basic test instrument for the beginner, an oscilloscope is widely used in phase relationships, electronic communication circuits, and other specialized industrial applications. An oscilloscope is shown in **Figure 13-8.**

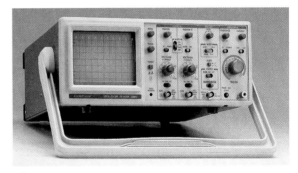

Figure 13-8.
An oscilloscope used in many industrial applications. (Knight Electronics)

Quiz–Chapter 13

Write your answers to these questions on a separate sheet of paper. Do *not* write in this book.

1. On graph paper, construct a sine wave. Show the rotating vector and plot points at 30 degree intervals of rotation.
2. A dc generator uses a commutator, but the ac generator has _____.
3. A vector represents the _____ and _____ of a force.
4. Frequency of an alternating waveform is measured in _____.
5. A certain electric current has a frequency of 400 Hz. What is the time interval or period of one cycle?
6. What is the effective value of a 150 peak voltage?
7. What is the peak voltage of the electricity in your home that is rated at 120 volts?
8. Meters reading ac values are usually indicating _____ voltage and current values.
9. On graph paper draw a single-phase wave, a two-phase wave, and a three-phase wave.
10. What is the rms of an average voltage of 90 volts?

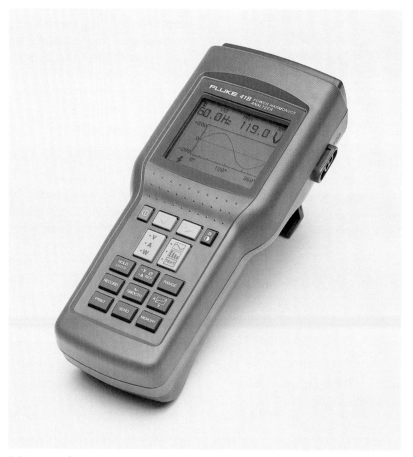

Many modern test instruments, like this power analyzer, can display their results graphically. (Fluke Corp.)

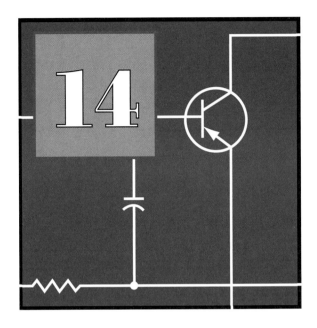

Capacitance

Objectives

After studying this chapter, you will be able to answer these questions:
1. What is a capacitor?
2. What is the nature of capacitance in a circuit?
3. What is the effect of capacitance in a circuit?
4. What is capacitive reactance?

Important Words and Terms

The following words and terms are key concepts in this chapter. Look for them as you read this chapter.

capacitance	fixed capacitor
capacitive reactance (X_C)	impedance (Z)
capacitor	microfarad
ceramic capacitor	picofarad
dielectric	rotor
dielectric constant	stator
electrolytic capacitor	time constant
farad	variable capacitor
	working voltage

By definition, *capacitance* is that property of a circuit that opposes any change in voltage within a circuit. Capacitors provide this function by their ability to store energy in the form of an electric charge.

Capacitors

Figure 14-1 shows two metal plates. These plates are not touching each other. There is air insulation between each of the plates. This setup creates a *capacitor*. A capacitor has the ability to store an electric charge. One of the metal plates is connected to the negative side of a battery and the other metal plate to the positive side. The switch permits the circuit to be turned on or off. A schematic diagram of the circuit is also shown in Figure 14-1.

When the switch is closed, electrons from the negative terminal move toward plate A. Because plate A is very close to plate B, the added electrons from the battery on A repel the electrons already on plate B. These repelled electrons are attracted to the positive battery terminal. Remember, opposite charges attract.

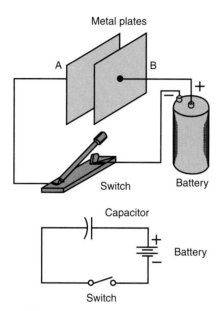

Figure 14-1.
The simple capacitor connected to a battery and the schematic diagram.

Consequently, plate A becomes negative (excess of electrons) and plate B becomes positive (fewer electrons). The capacitor is now *charged* to an equal but opposite voltage of the battery. The capacitor accepts a charge and becomes a voltage source in its own right.

The charging action of a capacitor can be studied in **Figure 14-2**. It is important to realize that current flows only during the charging action. When the capacitor is fully charged, the circuit current drops to zero. No current actually flowed through the capacitor. This is because the air gap between the plates acts as an insulator.

The Capacitive Circuit

As stated earlier, capacitance is defined as that property of a circuit that opposes a change in voltage. The capacitor is so named because it has the ability to store an electric charge. In **Figure 14-3**, it is interesting to observe the current and voltage during capacitor charging.

Safety Suggestion!

You have just learned a very important safety concept. Capacitors can, and do, store an electrical charge. The charge can be compared to any other power source—batteries or power supplies. Always consider a capacitor to be charged. For this reason, make sure a capacitor is discharged before you work with it. Be very careful as a capacitor's value of capacitance (stored charge) increases. The safest way to discharge a capacitor is with two clip leads and a low-value (about 100 ohms), high-wattage (one watt or higher) resistor. The clip leads should be connected to the resistor and (with the power source removed from the capacitor) one clip and then the other should be attached to the capacitor that needs to be discharged. This method can be called using a *bleeder* resistor. This resistor bleeds off the stored charge. Note: Special high-capacitance, high-voltage capacitors may need special discharge procedures. Always refer to manufacturer recommendations when discharging these types of capacitors.

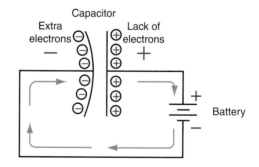

Figure 14-2.
Charging current flows for a fraction of time until capacitor becomes charged. The time of charge depends upon the resistance and capacitance in the circuit.

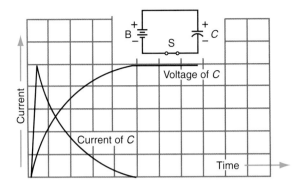

Figure 14-3.
As the capacitor becomes charged, the current drops to zero.

When the switch is initially closed and the capacitor is completely discharged, maximum current starts to flow. As C charges, the charging current becomes less and less until it is zero and C is fully charged. When current is maximum, the charge on C is minimum. When the charge of C is maximum, the current is zero. In a capacitive circuit, the voltage and current are 90 degrees out of phase. The current leads the capacitor voltage by an angle of 90 degrees (in a pure capacitive circuit) or less (if there is resistance also in the circuit).

The Farad

Capacitance is measured in *farads* in honor of the British scientist Michael Faraday. A capacitor that stores a charge of one coulomb of electrons when one volt is applied is said to have one farad of capacitance. The coulomb is composed of a very large number of electrons (6.24×10^{18} electrons). Generally speaking, farads and even millifarads are too great a unit for most electrical and electronic applications. In practical applications of capacitors, they will have values in *microfarads* and *picofarads*.

1 microfarad = 1 µF = one millionth of a farad

1 picofarad = 1 pF = one millionth of one millionth of a farad

Occasionally, you may see capacitor values listed as micro-microfarads or µµF. This unit of measure is the same as pF. In older capacitors, the method of labeling them was with a lowercase "m." Do not be confused with this lowercase m meaning millifarad. A lowercase m in capacitor labeling means microfarads, such as mF = µF, or, mmF = µµF = pF. Also, do not be confused if you see the letter "d" following the capital "F." This is also an older designation for farads—F = Fd.

The chart shown in **Figure 14-4** shows the base unit of capacitance, the farad. It also shows much smaller units of capacitance used in typical circuits. Note that microfarads, µF, are 10^{-6} power of ten, while picofarads, pF, are 10^{-12} power of ten. Indeed, values of capacitance are relatively small. Remember, they are still able to store a tremendous amount of electrical charge in a very small package.

Working Voltage

The insulation used between the plates of a capacitor and the distance between the

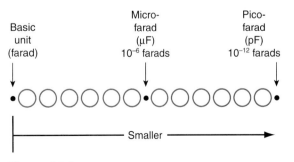

Figure 14-4.
Prefixes for capacitance units. (Note that millifarads, 10^{-3}, are generally not used.)

plates determines what maximum voltage can be applied to a capacitor without electricity arcing (sparking) between the plates. This arcing will make the capacitor less effective or even destroy it.

This maximum voltage that can be safely applied to a capacitor is called the *working voltage* and will be stated on the capacitor. It should not be exceeded. However, it is possible to use a higher rated working voltage capacitor in a lower voltage circuit. For example, it would be possible to use a 100 volt capacitor in a 50 volt circuit. It is always best to replace any component, capacitors included, with the same values as the device that was removed.

Labeling Capacitors

We have discussed that capacitors have a value of farads (capacitance), usually expressed in the much smaller units of microfarads (µF) or picofarads (pF). Additionally, capacitors have a working voltage rating. This is the maximum ac or dc voltage value that can be safely applied to the capacitor without damaging the insulation between the plates. Capacitors are also rated by their tolerance value. Similar to that with resistors, this is the range above and below the listed value of capacitance that the capacitor will be. If the capacitor is within this range, it is still within tolerance (proper working value). Typical capacitor ratings appear on the component written like this:

47 µF, 50 VDC, 10%

This means the capacitor value is 47 µF (value of capacitance), its working voltage is 50 volts dc (it must be operated in circuits of 50 volts dc or less), and its tolerance is 10%. The actual value of capacitance can be 10% below 47 µF:

47 µF − 4.7 µF = 42.3 µF

or 10% above 47 µF:

47 µF + 4.7 µF = 51.7 µF.

In smaller-sized capacitors, different methods must be used to label the device. This is because of the actual physical size of the device. Capacitors can be very tiny and all required information will not fit on them. In the past, a color code technique was used. It was very similar to the resistor color code method. This method has been largely replaced by a numbering system. Small-sized capacitors, like a ceramic type, may only have three numbers printed on it. For example:

103

In this example, the 1, or first digit, is a value of capacitance. So is the second digit, 0, in this component. The third digit, 3, represents the number of zeros to add to the first and second digit. The 103 gives this capacitor a value is 10,000. Because these capacitors are all relatively small, their unit abbreviation is assumed to be picofarads. A device labeled 103, is 10,000 pF, or 0.01 µF.

Factors that Determine Capacitance

Some of the factors that determine capacitance have already been mentioned. Naturally larger plates (greater area) will store more electrons. The distance between the plates is a factor since closer plates cause a greater density of electrostatic fields. An electrostatic field is created by excess electrons on one plate, and a shortage of electrons on the opposite plate. The area between the plates becomes an electrostatic field.

The kind of insulation used between the plates also varies a capacitor's ability to concentrate the electrostatic field. Generally, this insulation is referred to as the *dielectric*. The *dielectric constant* is a number that compares various insulators or dielectrics to that of air. Air or vacuum has been assigned a value of 1. Materials that exhibit a better insulation (or dielectric) characteristic than air have higher numbers. Higher numbers mean a larger capacitance. Mica has a dielectric constant of 5. Mica is a five times better dielectric than air. Waxed paper has a dielectric constant of 3, only three times better than air.

Summarizing these effects, capacitance varies:
1. Directly with plate area.
2. Directly with dielectric constant.
3. Inversely with distance between plates or thickness of dielectric.

Additionally, temperature also affects capacitance. However, compared to the three listed factors, temperature variation has only a slight effect on capacitance. There are a number of devices that test capacitors. You can also test to see if a capacitor is faulty using the ohms setting of a standard multimeter, **Figure 14-5.** The ohms value given should rise to infinity as the capacitor charges. Be careful to note capacitor polarity!

Kinds of Capacitors

Capacitors are manufactured in a wide variety of types, sizes, and values. *Fixed capacitors* only have one value of capacitance. *Variable capacitors* have a range of possible capacitances. In **Figure 14-6**, a variable capacitor is illustrated. It can be used in tuning circuits of radios as well as other applications. Note the symbol for a variable capacitor. It is the symbol of a standard capacitor with an arrow crossing through it.

In Figure 14-6, the fixed plates of the variable capacitor are called the *stator* and the rotating plates are called the *rotor*. The insulation or dielectric in this case is air.

A common type of fixed capacitor consists of two layers of metal foil separated by waxed paper. This is rolled up into a cylinder and usually encapsulated in plastic. Wire leads connected to the foil

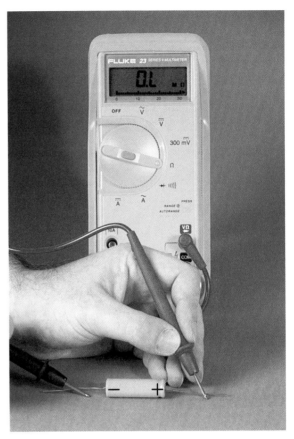

Figure 14-5.
Testing a capacitor with a DMM.

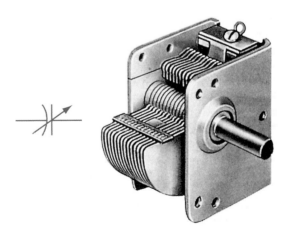

Figure 14-6.
An air dielectric variable capacitor shown with the symbol for a variable capacitor.

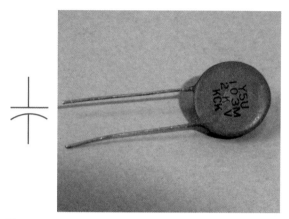

Figure 14-7.
Ceramic capacitor.

> **Safety Suggestion!**
>
> The electrolytic or tantalum capacitor must never be placed in the circuit with the wrong (reverse) polarity. Doing so will ruin the oxide chemical coating inside the capacitor and cause a short circuit. Always be sure to install electrolytic capacitors carefully and note the polarity signs. If an electrolytic or tantalum capacitor has been put in the circuit incorrectly, remove it and discard it. If the oxide coating was not totally destroyed, its life expectancy has certainly been reduced, and it may not work reliably.

plates form the external connections to the capacitor. This type is called a "paper capacitor." The plastic coated types are rigid, less fragile, and can be used at higher temperatures. However, this type of capacitor is being used less in electronic applications because it does not function as well as other types.

Another group of capacitors are the *ceramic capacitors* illustrated in **Figure 14-7**. These capacitors are made by using ceramic as an insulator and a silver deposit on each side for the plates. These capacitors are used for small values of capacitance. They work well at high voltages.

The *electrolytic capacitor* is used when a relatively large amount of capacitance is needed in a small space. It is formed by chemical action. The dielectric is a very thin coating of oxide between the two plates. You will find these metal cans in plastic wrapping as shown in **Figure 14-8**. Note that some capacitor cans may contain two or more capacitors.

One very important point you must remember about the electrolytic capacitor is that one end is positive (+) and the other negative (–). You must observe this polarity when connecting electrolytic capacitors in your circuits. The symbol for these capacitors is the standard capacitor symbol with positive and negative marks. These polarity marks will tell you it is an electrolytic or polarized capacitor.

Another type of polarized capacitor is the tantalum capacitor. Like the electrolytic, the tantalum capacitor has a relatively large amount of capacitance. However, the

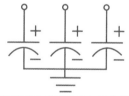

Figure 14-8.
Selection of electrolytic capacitors and symbol for three-section capacitors.

chemical used in this device takes up even less space (surface area) than the electrolytic capacitor. Again, polarity must be applied properly, or the device will be destroyed.

RC Time Constants

Earlier in this chapter the time required to charge a capacitor was discussed. If there is resistance in series with the capacitor, the charging time period will be extended because the resistor limits or opposes the current. The time required to charge a capacitor to about 63 percent of the maximum voltage in an RC circuit is called the *time constant* of the circuit. Time constants can be found using the formula:

$$\text{time (in seconds)} = R \text{ (in ohms)} \times C \text{ (in farads)}$$

Five time constant periods are required to fully charge or discharge a capacitor. In **Figure 14-9,** an RC circuit is drawn and the time constant computed. This circuit has a time constant of one second. Five seconds are required for C to become fully charged.

Math Manipulation!

Practice calculating various values of R and C for the circuit shown in Figure 14-9. Note that the time constant formula has a direct relationship between resistance and capacitance. If either value increases, the time constant increases. Likewise, if either value decreases, the time constant value decreases as well.

Capacitive Reactance

In a dc circuit, a capacitor charges to the value of the applied voltage and the current becomes zero. An infinite amount of

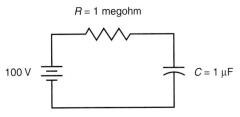

Time constant = $(1 \times 10^6) \times (1 \times 10^{-6})$ = 1 second

Figure 14-9.
In this RC circuit, C will charge to about 63 volts in one second. At the end of five seconds, C will be charged to about 100 volts.

resistance (∞) also produces zero current, so we can conclude that a capacitor *blocks* a dc current. This is called a dc blocking capacitor.

However, look what happens when an ac voltage is connected to the capacitor. This is illustrated in **Figure 14-10**. During the positive half cycle, current flows in one direction. During the negative half cycle, it flows in the opposite direction. Current is always flowing in one direction or the other.

How much current is there? That depends upon the capacitance value of the capacitor and the frequency of the applied

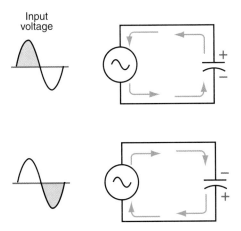

Figure 14-10.
Current flows in a capacitive circuit when an ac voltage is applied. The amount of current depends on the reactance of C in ohms.

voltage. A capacitor offers opposition to the flow of alternating current. This opposition is called *capacitive reactance.* Its symbol is X_C. This opposition to ac current is expressed in ohms. Realize that the capacitor is not truly a resistor, however, it *reacts* to the flow of ac and decreases it. Remember that current does not technically flow through the capacitor. The current is influenced to flow by the power source.

Capacitive reactance becomes very important in various amplifier and filtering circuits. For example, if a great deal of bass is desired in an amplifier circuit, filters can emphasize low frequencies (bass) and de-emphasize high frequencies (treble). De-emphasis of various frequencies can be accomplished by capacitors and their capacitive reactance, their opposition to alternating current.

The capacitive reactance of a capacitor can be computed using the formula:

$$X_C = \frac{1}{2\pi f C}$$

where X_C equals the capacitive reactance in ohms, f equals the frequency of applied voltage in hertz, and C equals the capacitance in farads. π is approximately equal to 3.14.

Impedance

In a circuit containing both resistance and capacitance, there is a net total effect of opposition to the flow of alternating current. The total opposition of R and C cannot be added directly together since the capacitance causes the current to lead the voltage and the current and voltage in a resistor are always in phase. Therefore, a vector addition must be made to discover the resultant opposition. This resultant is called *impedance.* Impedance uses the letter symbol **Z**. Like resistance and reactance, it is expressed in ohms.

Refer to the diagram in **Figure 14-11** in which a vector diagram and circuit are calculated. The vector diagram of Figure 14-11 graphically shows us a theorem that is applied in geometry. The Pythagorean theorem states:

$$a^2 = b^2 + c^2$$

where a, b, and c are the sides of a right triangle as shown. Taking the square root of both sides of the equation:

$$a = \sqrt{(b^2 + c^2)}$$

This formula can be used to add resistance and capacitive reactance by rewriting the Pythagorean theorem as:

$$\text{impedance} = \sqrt{\text{resistance}^2 + \text{capacitive reactance}^2}$$

or

$$Z = \sqrt{(R^2 + X_C^2)}$$

In this way, by using a right triangle (in the vector diagram) and the Pythagorean theorem, the actual value of impedance (in ohms) can be calculated.

Math Manipulation!

As you review the circuit shown in Figure 14-11 and its vector diagram, consider the formula for Z, the impedance. The impedance of the circuit is equal to the square root of the resistance squared added to the capacitive reactance squared. Using a calculator, it is best to determine individual

Chapter 14   Capacitance

values for resistance squared (the value of resistance multiplied by itself) and capacitive reactance squared (the value of capacitive reactance multiplied by itself). Then, after taking the sum of these values, compute the square root of that quantity. Writing down these intermediate steps and values will make it easier to solve this problem without error.

In the following chapter, we will discuss the interaction of capacitive reactance and inductive reactance. We will also discuss the term *resonance* and its effect once we learn about inductors.

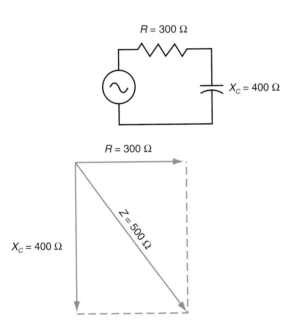

Figure 14-11.
A series RC circuit has resistance of 300 Ω and X_C of 400 Ω. Its resultant impedance Z is 500 Ω.

History Hit!

Pythagoras (c. 560–480 BC)

Pythagoras was a Greek mathematician and astronomer. He discovered the theorem based upon right angle triangles.

Quiz–Chapter 14

Write your answers to these questions on a separate sheet of paper. Do *not* write in this book.

1. What is the unit of measurement of capacitance?
2. Draw the symbol for a fixed capacitor.
3. Draw the symbol for a variable capacitor.
4. Do electrons repel or attract each other?
5. When selecting a capacitor, what else should you consider besides its value in farads?
6. Variable capacitors have fixed plates called the _____ and rotating plates called the _____. The insulation or dielectric is _____.
7. What is the name of the common type of fixed capacitor that consists of a roll of two layers of metal foil separated by waxed paper?
8. What term is used to describe the opposition to alternating current flow through a capacitor?
9. X_C is a symbol for what property of an electronic circuit?

120 Electricity

Training in electronics opens a wide range of careers. (Fluke Corp.)

Inductance

Objectives

After studying this chapter, you will be able to answer these questions:
1. What is an inductor?
2. What is inductance in a circuit?
3. What is inductive reactance?
4. What is resonance?

Important Words and Terms

The following words and terms are key concepts in this chapter. Look for them as you read this chapter.

counterelectromotive force (cemf)
henry
inductance
inductive reactance (X_L)
inductor
quality (Q)
resonance
resonant frequency
self inductance

In the last chapter, capacitors were defined as devices that opposed any change in voltage within a circuit. The device to be covered in this chapter is the inductor. An *inductor* provides inductance to a circuit. By definition, *inductance* is that property of a circuit that opposes any change in current in a circuit.

Let's explore the meaning of this definition and show how it can be proved. A review of *Magnetism,* Chapter 10, may be necessary. Associated with a conductor carrying a current is a magnetic field. If the conductor is wound into a coil around a core, the magnetic fields around each turn of the coil join and reinforce each other so that the coil becomes an electromagnet with a definite north–south polarity. Existing in space around the coil are many invisible magnetic lines of force, or flux lines. It could be a large and strong magnetic field or it could be a weak magnetic field depending upon the value of the current through the coil.

Self Inductance

In *Direct Current Generators,* Chapter 12, the discoveries made by Faraday were used to explain the action necessary for the generation of electricity. You will recall that three conditions must exist.

1. A magnetic field must be present.
2. A conductor (a wire or coil) must be available.
3. There must be relative motion between the field and the conductor. Either the conductor must move or the field must move.

In **Figure 15-1,** a simple coil of wire formed on an iron core has been connected to a dc power source. The switch is open. There is no current flowing. There is no magnetic field.

In **Figure 15-2,** the switch has been closed and a current flows in the circuit (coil) producing a magnetic field around the wire and core. The greater the current, the stronger the magnetic field. We can conclude that:
1. No current—no magnetic field.
2. Increasing current—increasing magnetic field.
3. Maximum current—maximum magnetic field.

This switching on of the circuit creates a *moving magnetic field*. The magnetic field expands outwardly as the current increases; it contracts or moves inwardly as the current decreases. This moving field cuts across the wires that form the coil. This moving field produces a voltage called **counterelectromotive force (cemf),** which opposes the source or applied voltage. Counterelectromotive force, or cemf, opposes any increase or decrease of current

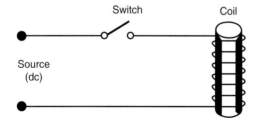

Figure 15-1.
A coil wound on an iron core is connected to a dc power source. The switch is open.

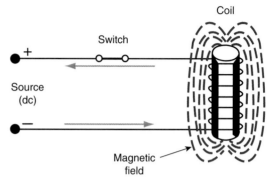

Figure 15-2.
When the switch is closed, a current flows in the coil and produces a magnetic field.

through the coil. Counterelectromotive force can be called an *induced* voltage resulting from the moving magnetic field across the wires of the coil. This phenomenon is termed *self inductance.*

It is interesting to note that if a steady direct current of any value, even a thousand amperes, were flowing through the coil it would have no cemf. However, if the current changes, even as little as one microampere, the magnetic field would vary and a cemf would develop that would oppose the current change.

The Henry

A coil is called an inductor. If the rate of change of one ampere per second produces a cemf of one volt, the coil is described as having an inductance of one *henry.* This unit of measurement honors the memory of Joseph Henry.

Many coils, with and without cores, are used in electricity and electronics. They have values in henrys, millihenrys, and microhenrys depending upon their application. The chart shown in **Figure 15-3** depicts the base unit of inductance.

Examples of these coils and the inductor symbol are illustrated in **Figure 15-4.**

History Hit!

Joseph Henry (1797–1878)

Henry was a physicist from the United States who pioneered electromagnetism and its uses. He taught at the College of New Jersey—which later became known as Princeton University. Henry discovered the conversion of magnetism into electricity, and for this reason the unit of inductance is named in his honor. Additionally, Joseph Henry was the first director of the Smithsonian Institution, named after James Smithson, an English scientist, who contributed a fortune to establish it.

Labeling Inductors

We have discussed the term used to measure the inductance of inductors—the henry. Usually, inductor values are in the range of millihenrys or microhenrys. Large inductors, like those used in power supplies, can have values in the range of 1 to 10 henrys.

Inductors have other ratings as well. These include the overall dc resistance of the coil of wire within the inductor, the maximum amount of current it can handle, and its overall tolerance. The tolerance of the inductor determines how much above or below its listed value the component can be and still be within its acceptable limit. A typical inductor rating may be:

45 mH, 100 Ω, 50 mA, 10 V, 5%

The inductor value is equal to 45 millihenrys, the dc resistance is 100 ohms, maximum current in the device is 50 milliamps, the maximum voltage that can be applied to the device is 10 volts, and the tolerance is 5%. This tolerance level means the device can be 2.25 mH above or below the listed value and still be within proper functioning limits (42.75 mH–47.25 mH).

One additional rating of an inductor is the *quality* rating or *Q*. Quality, or *Q*, of an inductor is determined by the ratio of the X_L, inductive reactance, to *R*, resistance. In formula form:

$$Q = \frac{X_L}{R}$$

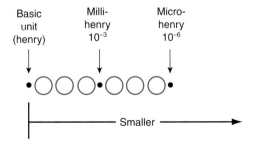

Figure 15-3.
Prefixes for inductance units.

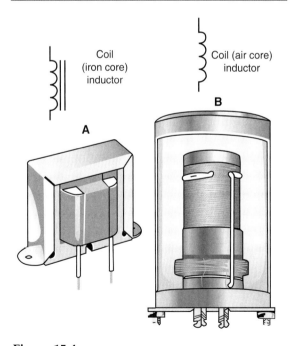

Figure 15-4.
Symbols (top) for coils and examples of coils used in electronic circuits. A—Choke coil used in a power supply. B—An antenna coil inductor.

In smaller sized inductors, different methods must be used to label the device. All the ratings will not fit on the component body. In the past, a color code technique was used. Similar to resistor color coding, this method can be found in use today. In fact, in some cases it can be difficult telling the difference between an inductor and resistor without looking carefully. Sometimes, one can only tell the difference by looking at the circuit schematic. Often, an inductance meter is used to accurately determine the value of inductance.

Coil Action in a DC Circuit

Consider the resistance only circuit in **Figure 15-5.** When the switch is closed, the current rises to a maximum value almost instantaneously and remains at a fixed value as determined by Ohm's law (voltage divided by resistance).

In **Figure 15-6,** an inductor is used instead of the resistor. In this case, the current does not rise instantaneously. This is because the rising current induces a counter-electromotive force (cemf) that opposes the rising current. The time delay of the current reaching its peak level depends upon the inductance of the coil and any resistance that may be in the circuit. The wire used in the coil will have some resistance. This delay is referred to as the *time constant* of the circuit. Mathematically:

$$\text{time (in seconds)} = \frac{L}{R}$$

where L is the inductance in henrys and R is the resistance in ohms.

This is the time required for the current to rise to approximately 63 percent of its final value. At the end of *five* time constant periods, the current is considered to be at its maximum or steady-state value. It is important to remember that inductance appears *only* when there is a change in current. When the current has risen to its steady-state (unchanging) value, there is no inductance and the current flowing in the circuit is limited only by the dc resistance of the circuit.

Inductive Reactance

When an inductor or coil is used in an ac circuit, a new situation develops. Assuming a sine wave ac voltage is applied to the circuit, the current rises and falls and reverses direction in step with the applied

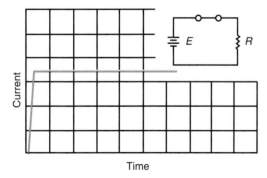

Figure 15-5.
In a resistive circuit, the current rises instantaneously to some fixed value depending on the voltage applied and the value of resistance.

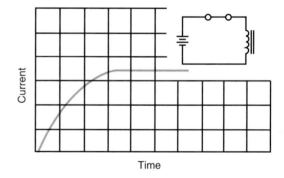

Figure 15-6.
In an inductive circuit, the current rise is delayed because of the cemf induced in the coil by a change in the applied current.

voltage and at the same frequency as the applied voltage. Since a constantly changing current would produce a continuous cemf, the current never rises to its maximum value as it did in a dc (steady-state) circuit. The value of the ac current is determined by the *inductance of the coil* and the *frequency* of the applied ac voltage. This opposition to the flow of ac current is called **inductive reactance**, and it is expressed in ohms just like dc resistance. The symbol for reactance is X, and since it is caused by a coil or inductor, it is labeled X_L for inductive reactance. The formula appears as:

$$X_L = 2\pi f L$$

where X_L is the inductive reactance in ohms, f is the frequency of applied voltage in hertz, L is the inductance of the coil in henrys, and π is approximately equal to 3.14.

Impedance

Considering the previous discussion, the inductance and the cemf prevent the current from rising and falling in step with the applied voltage. There is a delay and the current will kind of "drag its feet" all the time. In a theoretically pure inductive circuit, the current will, in fact, lag behind the applied voltage by 90 degrees.

In a purely resistive circuit, the voltage and current rise and fall together in step or in phase. Any circuit that contains both inductance and resistance causes the current to lag by some angle less than 90 degrees. To make this statement in another way, there are two oppositions to the flow of current: the inductive reactance (X_L) and the dc resistance (R) in ohms. Since the opposition by X_L is 90 degrees out of phase with R in ohms, they cannot be added directly together. The resultant of the two oppositions must be found. In **Figure 15-7,** an example of this vector addition is illustrated. The resultant, or total opposition to ac current from X_L and R, is called the impedance of the circuit. Its symbol is Z and it is also expressed in ohms.

The vector diagram of Figure 15-7 graphically shows us the use of the Pythagorean theorem, introduced in Chapter 14. Similar to capacitive reactance, X_C, total opposition to ac current flow in an inductive circuit is called impedance or Z.

An application of the Pythagorean theorem in geometry tells us that:

$$Z^2 = R^2 + X_L^2 \text{ or } Z = \sqrt{(R^2 + X_L^2)}$$

Combining X_L and X_C in a Circuit

A thorough understanding of the values of inductive reactance (X_L) and capacitive reactance (X_C, as described in

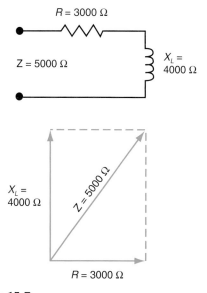

Figure 15-7.
The impedance of a series RL circuit is found by adding X_L and R vectorially since X_L is 90 degrees out of phase with R.

Web Wanderings!

http://NIST.gov/

The National Institute of Standards and Technology (NIST) works with industry to develop and apply technology, measurements, and standards. On their Web site you will find standards information and links to scientific and technical databases and publications for industry, researchers, and the general public.

Chapter 14) can be made by examining them as the frequency change. Compare the two graphs in **Figure 15-8** and make these conclusions:

- X_L increases in resistance with higher frequencies.
- X_C decreases in resistance with higher frequencies.

At zero frequency or dc:
X_L = 0 ohms.
X_C = infinity (∞) ohms.

At high frequencies:
- X_L increases linearly (a straight line increase).
- X_C approaches zero resistance in a nonlinear fashion (curved line decrease).

Resonance

Resonance in an inductive/capacitive circuit is when X_L equals X_C. In a circuit containing R (resistance), L (inductance), C (capacitance), as in Figure 15-8, at one single frequency X_L is equal to X_C. At this point, X_L cancels out X_C since they are 180 degrees opposite to each other. The impedance of the circuit is equal to only R.

$$Z = R \text{ (at resonance)}$$

The particular frequency at which this occurs is known as the *resonant frequency* of the circuit. Mathematically, the resonant frequency equals:

$$f_r = \frac{1}{2\pi\sqrt{LC}}$$

where f_r is the resonant frequency, π is approximately 3.14, L is the inductance in henrys, and C is the capacitance in farads.

This formula seems complex at first. However, it combines the formulas for finding inductance and capacitance. Inductance and capacitance are multiplied together and then the square root of the value is taken. This value is then multiplied by 2π (3.14 multiplied by 2 = 6.28). The value of π is used because of the method of generation of the sine waveform.

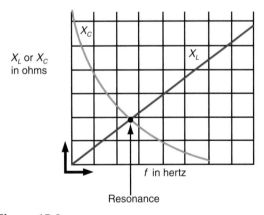

Figure 15-8.
On the graph is plotted the value in resistance (ohms) of X_C and X_L as frequency is increased from zero frequency or dc. Resonance is the frequency where X_C and X_L cross each other and become equal in value. This single point is where the impedance is at its minimum.

Your more advanced studies will show how resonant circuits play a major role in radio communications, television, cellular phones and many other electronic communication applications.

Quiz–Chapter 15

Write your answers to these questions on a separate sheet of paper. Do *not* write in this book.

1. Draw the symbol for an inductor.
2. What is the letter symbol for inductance?
3. What is the symbol for inductive reactance?
4. What is the unit of measurement for inductance?
5. What is the formula for inductive reactance?
6. Define inductance.
7. In order to induce current flow in a conductor, there must be relative motion between the conductor and a _____.
8. Draw a schematic diagram for an inductor (*L*) in series with a resistor (*R*), connected to a battery through a switch (S).
9. Can an inductor be used to filter out variations in current? Explain.
10. Can direct current resistance and inductive reactance be added together like series resistors? Explain.
11. After the current has built up in an inductive circuit and there is no further change, the inductance appears only as resistance in a direct current circuit. Explain.
12. Draw a graph showing the difference between the rise in current in an inductive circuit and a pure resistive circuit.

This technician is using induction to take readings. By measuring the magnetic field around a conductor, you can take accurate current readings without breaking the circuit. (Fluke Corp.)

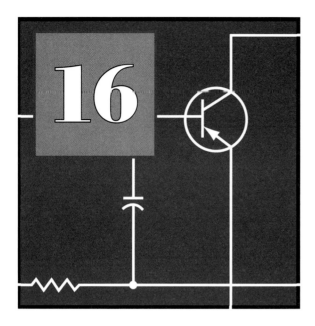

Transformers

Objectives

After studying this chapter, you will be able to answer these questions:
1. What is the theory and operation of induction coils?
2. What is the theory of the operation of a transformer?
3. What are some common applications of the transformer?

Important Words and Terms

The following words and terms are key concepts in this chapter. Look for them as you read this chapter.

autotransformer
core-type transformer
coupling
induced
isolation transformer
primary winding
secondary winding
shell-type transformer
step-down transformer
step-up transformer
transformer
turns ratio
unity coupling

A *transformer* is a device used to transform (change) and/or transfer electrical power from one circuit to another.

To understand how this can be done, let's review the lessons on the generator. You will remember the experiments of Michael Faraday, when he discovered the principles by which an electric current could be produced from a magnetic field. In a simple generator, the magnetic field is created by the field windings and magnets. The rotating armature supplied the conductors and the relative motion between the field and conductors. Another way to produce a similar effect is to hold the conductor stationary and move the magnetic field. This technique produces the same result.

In the study of the transformer, we will first consider an ordinary electromagnet, to which is connected a source of alternating current. As the alternating current rises to maximum flow in one direction, the magnetic field expands outward to maximum strength. As current decreases, the magnetic field decreases or collapses until it reaches the zero point. The alternating current then rises to maximum in the opposite direction, creating a magnetic field with opposite polarity. An alternating current, therefore, produces a moving magnetic field, ever changing in its strength and polarity at the same frequency as the applied alternating current.

If another conductor or coil is placed close to this varying magnetic field, a current will be *induced* to flow in this coil. An induced current is created by the varying magnetic field passing through the conductor. The magnetic lines of force of the electromagnet cutting across this conductor or coil induce current in it. **Figure 16-1** is a schematic diagram of such a device.

Coil A produces the ever-changing magnetic field. It is called the *primary* (or input) *winding.* The induced current flows in coil B. It is known as the *secondary* (or output) *winding.* The number of lines of magnetic flux that link the primary coil and secondary coil together depend, among other things, upon the distance between them. This is called *coupling.* Normally, the secondary windings are so arranged that the maximum number of magnetic flux lines will cut through it. If all the magnetic flux lines from the primary winding were coupled to the secondary coil, this would be called *unity coupling.* Unity coupling, though ideal, is never reached.

> **History Hit!**
>
> Heinrich Friedrich Emil Lenz (pronounced lents) (1804–1865)
> Lenz was a Russian physicist who made many important discoveries concerning sea temperature and salinity. However, he is best known for his efforts centered around electromagnetism.

Lenz's Law

It is important to remember that when a current flows and a magnetic field is produced in the primary, *the induced current flow in the secondary is in the opposite direction and creates a magnetic field of opposite polarity.* In other words, the induced magnetic field opposes the primary magnetic field. This is known as the Lenz's law.

Transformer Losses

The transformer is a very efficient device. However, some losses during the transfer of electrical energy (power) do occur. These losses were first covered in Chapter 12, in the discussion of generators.
1. *Copper losses*—Created by the resistance of the copper wire in the primary and secondary windings.
2. *Eddy current losses*—These losses occur when tiny circular currents are generated in the conductive iron core of the transformer. They are easily overcome by making the transformer core out of many thin sheets of iron that are insulated from each other. These thin sheets are called *laminations.*

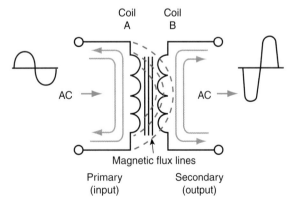

Figure 16-1.
The schematic diagram of a transformer.

3. *Hysteresis loss*—Here, losses are caused by the heating effect of molecular friction in the iron core. This is due to the continuous change in the magnetic field resulting from the nature of the alternating current applied. This is overcome by using special silicon steel and heat-treating processes for core materials.

Transformer Construction

Two common methods of transformer construction are used. In the first case, the primary and secondary windings are wound on each side of a laminated core in the form of a rectangle, **Figure 16-2.**

This is a *core-type transformer.* You might assume that the primary was wound on one side and the secondary on the other. Such is not the case however. Part of each winding is wound on each side to overcome coupling losses.

A second method of construction produces what is known as the *shell-type transformer.* It has the primary and secondary windings together on the center section of a laminated core, **Figure 16-3.**

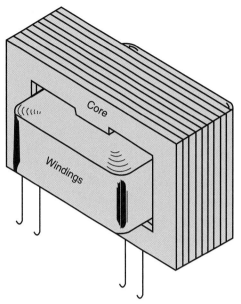

Figure 16-3.
A shell-type transformer.

These two methods of construction are easily distinguishable. You need simply remember that: *In the core-type transformer the windings are around the core, and in the shell-type transformer the core is around the windings.* Each type of transformer construction has special applications in the distribution and conversion of electrical power.

Turns Ratio

The *turns ratio* of a transformer lists the actual number of turns of wire on the primary and secondary coil. The ratio between the number of turns of wire on the primary and on the secondary depicts the operation of the transformer.

For example, if there are an equal number of turns in both the primary and secondary coil windings, the magnetic flux cuts across both windings and induces a voltage in the secondary that is the same as the voltage applied to the primary. Such a transformer has applications as an isolation transformer. An *isolation transformer* provides a physical separation between a

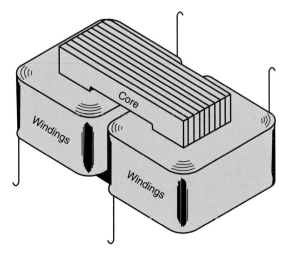

Figure 16-2.
A core-type transformer.

circuit and the applied ac power that is magnetically coupled. Technicians use the isolation transformer as a protective device against electric shocks and for expensive test equipment.

However, if there are twice as many secondary windings as primary windings, each line of magnetic flux of the primary cuts across two secondary windings. The induced voltage is then *twice* the applied voltage. Any combination of windings might be used and the secondary voltage varies in respect to the primary voltage as the relationship between the number of turns of wire in each. The formula appears as:

$$\frac{E_{in}}{E_{out}} = \frac{N_P}{N_S}$$

where, E_{in} is the applied voltage (primary), E_{out} is the voltage of secondary, N_P is the number of turns of wire in the primary, and N_S is the number of turns of wire in the secondary.

Math Manipulation!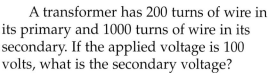

A transformer has 200 turns of wire in its primary and 1000 turns of wire in its secondary. If the applied voltage is 100 volts, what is the secondary voltage?

Substituting the known quantities in the formula, we have:

$$\frac{100 \text{ V}}{E_{out}} = \frac{200 \text{ turns}}{1000 \text{ turns}}$$

or

$$E_{out} = \frac{100{,}000}{200} = 500 \text{ V}$$

Safety Suggestion!

The isolation transformer is a standard device used by many technicians in the field. By means of separating (or isolating) the primary side of a power source to its secondary, a much more safer electrical environment exists. This is because the isolation transformer will prevent other potential paths of current in devices wired to the 120 volt ac line.

The relationship between the number of turns in the primary and secondary is known as the turns ratio. A transformer that raises the voltage is a *step-up transformer.* Conversely, a *step-down transformer* lowers the voltage.

Transformers can have several secondary windings to provide low and high voltages for the operation of various circuits, **Figure 16-4**. Transformers are also used as coupling devices for radio frequency signals.

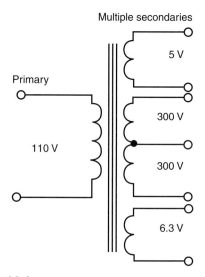

Figure 16-4.
The schematic diagram of a power transformer. The three vertical lines between the coils represent the laminations of the core.

Chapter 16 Transformers

Power Transmission

The following question is usually asked by the beginning student of electricity: "Why is alternating current almost universally used in home and industry in the United States?" The answer to the question is best described by the fact that electricity is more efficiently transmitted long distance by alternating current. In addition to there being lower electrical losses with alternating current, higher voltages decrease line (transmission) losses. The transformer can conveniently change voltage levels to aid in the transmission of power.

To better understand this concept, we will review some background information. A wire has a certain resistance to the flow of electricity, and a loss of power is the result of this resistance. Watt's law says that:

$$P = I^2 R$$

Because power loss increases as the *square of the current*, the loss becomes quite a factor when the current is increased. Compare these examples. With 10 amperes of current flowing through one ohm of resistance, the loss equals:

$$P = (10 \text{ A})^2 \times 1 \text{ } \Omega = 100 \text{ watts}$$

If the current were increased to twice its value, or 20 amps, the loss is:

$$P = (20 \text{ A})^2 \times 1 \text{ } \Omega = 400 \text{ watts}$$

or four times greater.

It is apparent that the current should be kept as low as possible to avoid power loss. But as power is the product of current times voltage, various combinations of I and E will produce the same power. Work through the following examples.

100 volts × 10 amperes = 1000 watts

1000 volts × 1 ampere = 1000 watts

10,000 volts × 0.1 ampere = 1000 watts

100,000 volts × 0.01 ampere = 1000 watts

If a large quantity of electrical energy is to be transmitted over a long distance, such as from a power plant in the country to a city a hundred miles away, the transformer can be used to overcome the power loss. At the power plant the voltage is stepped up to a very high value by transformers. The electricity is then transmitted to the city by high-tension power lines. Generally, a substation reduces these high voltages. At a distribution substation in the city, the voltage is again stepped down by transformers to lower voltages. A utility pole transformer or pad mounted transformer will again step the voltage down to the 120/220 volt ac power that is used in your home, **Figure 16-5**.

A sketch in **Figure 16-6** shows the sequence of events. Trace the current flow in each leg of the transmission circuit when 12,000 watts (12 kW) of power are transmitted from the power plant (point A) to your home (point D).

At the power plant generator:

$$I = \frac{P}{E} = \frac{12,000 \text{ watts}}{12,000 \text{ volts}} = 1 \text{ ampere}$$

Figure 16-5.
A pole utility transformer (left) and a pad mounted transformer (right).

134 Electricity

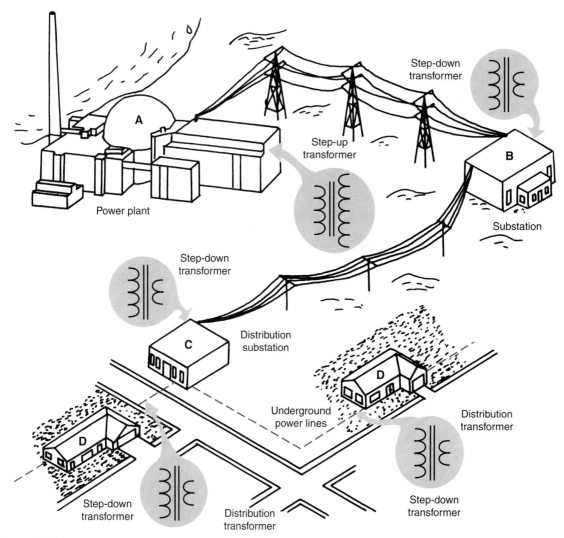

Figure 16-6.
The transmission of power from the power plant generator to your home.

The transformer at the power plant, then, steps up the voltage to 60,000 volts. At 60,000 volts, the current required for 12 kilowatts in the transmission line from the power plant (A) to the substation (B) is only 0.2 amps.

$$I = \frac{P}{E} = \frac{12{,}000 \text{ watts}}{60{,}000 \text{ volts}} = 0.2 \text{ amperes}$$

In the city at a substation (B), the voltage is stepped down to 1200 volts. The current for 12 kW of power at 1200 volts is 10 amperes.

$$I = \frac{P}{E} = \frac{12{,}000 \text{ watts}}{1200 \text{ volts}} = 10 \text{ amperes}$$

Your power may be stepped down two more times (points C and D) before entering your home. Before entering your home, the voltage has been stepped down to 120 volts and the current you are actually using is:

$$I = \frac{P}{E} = \frac{12{,}000 \text{ watts}}{120 \text{ volts}} = 100 \text{ amperes}$$

Summarizing

At point	$E \times I = P$	
A	$12{,}000 \times 1$	$= 12{,}000$ watts
A–B	$60{,}000 \times 0.2$	$= 12{,}000$ watts
B–C	1200×10	$= 12{,}000$ watts
D	120×100	$= 12{,}000$ watts

When you remember that the current carrying capacity of a wire depends upon its size, you will see some distinct advantages to stepping the voltage up before the transmission of electricity across the country. Smaller wires can be used. Small wires are more economical as well as being easier to install and maintain. Although smaller wires do add more resistance to the circuit, this power loss is more than offset by the savings made by using the small conductors.

Some observations can be made from the previous example of power transmission. A transformer cannot increase power. It is not a generating device. Power output is always equal to power input, neglecting any losses caused by the transformer.

> Power in the primary = Power in the secondary

and since power = $I \times E$:

> $I_{primary} \times E_{primary} = I_{secondary} \times E_{secondary}$

In a transformer the secondary voltage varies directly with the turns ratio. Restated the turns ratio formula from earlier in this chapter:

$$\frac{E_P}{E_S} = \frac{N_P}{N_S}$$

Therefore, in order to keep the power in the primary approximately equal to the power in the secondary, an increase in voltage must be accompanied with a *decrease* in current. A decrease in voltage must have an *increase* in current. With respect to current:

> $$\frac{I_P}{I_S} = \frac{N_S}{N_P}$$

Notice that current is inversely proportional to the turns ratio.

The Induction Coil

Frequently it is necessary to raise a dc voltage for use in electrical equipment. One example to illustrate this transformer action is the ignition system of the automobile. The battery has a dc voltage of 12 volts. In order to cause a dc spark to jump the gap of a spark plug, a voltage of some 20,000 volts is needed. In certain high energy ignition systems, upwards of 100,000 volts is produced.

To do this, the current in the primary windings of the transformer is made to vary by the opening and closing of a switch. This switch can be of several types. In the past, automobiles relied upon mechanical switches (what was known as the "points"). Today, these moving switches have been replaced with electronic switches. Sealed inside these electronic ignition system modules are switching transistors that can handle the rapid switching (on and off) of the primary of the high voltage coil.

Pulsating direct current in the circuit somewhat simulates an ac voltage and creates a varying magnetic field that induces a current to flow in the secondary windings. The transistor switch in the primary circuit (the ignition module) is actuated by an electronic control module (the car's computer). **Figure 16-7.** The transformer in this automotive circuit is a known as the *high voltage coil* (a step-up transformer).

Another variation of the induction coil is one that uses magnetism of the primary to open a set of contacts and break a circuit, see **Figure 16-8.** In this circuit, the current flowing in the primary makes an electromagnet that pulls the contacts open. When this occurs, the primary current falls to zero, the magnetism disappears, and the spring pulls the contacts together again. This causes current to flow again. The action goes on as long as there is an applied

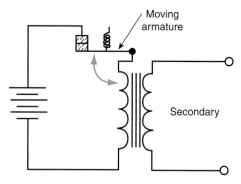

Figure 16-8.
A diagram of an induction coil. A magnetic field will pull the moving armature to the coil breaking the circuit. The spring will return the armature back into position and close the points. The contact opening and closing simulates ac voltage to the transformer.

voltage. A higher voltage is induced in the secondary because of the step up action of the transformer. This technique is often used in small gasoline engines such as found in push and riding lawn mowers.

This same circuit without the secondary winding is used as a door buzzer. When a bell striker is attached to the moving point, a doorbell can be made, **Figure 16-9.**

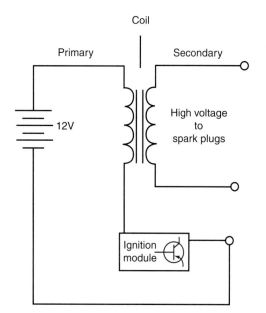

Figure 16-7.
A simplified diagram of an automotive ignition system. The ignition module "fools" the coil or transformer into thinking ac voltage is being applied.

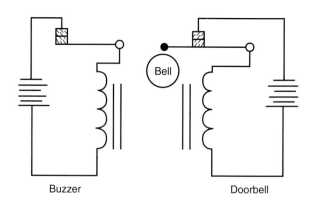

Figure 16-9.
Simplified diagram of a buzzer and doorbell. The physical motion of the contacts in the buzzer creates the sound heard.

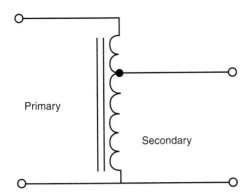

Figure 16-10.
The schematic diagram of a step-down autotransformer.

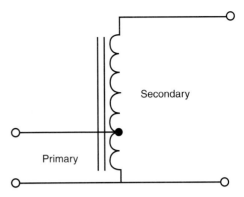

Figure 16-11.
The schematic diagram of a step-up autotransformer.

Circuits in *Project 5—Magnetic Relay* demonstrate how to connect the relay as a buzzer. It is easy to build and it will add to your knowledge of electricity and magnetism.

Autotransformer

It is possible to construct a transformer with only one winding. This winding serves as both the primary and the secondary. Refer to **Figure 16-10**. Such a device is called an *autotransformer*. The primary is connected across the input power supply. A tap on the primary is provided for one end of the secondary.

The ratio of the input and output voltage is in proportion to the turns ratio, however, the current in the secondary is low because the induced current flows in the opposite direction from the primary current. Hence the net current is the difference between the two currents.

An autotransformer is used in many television sets to produce the high voltage to operate the picture tube. In this application it would be a step-up transformer, **Figure 16-11.**

Quiz–Chapter 16

Write your answers to these questions on a separate sheet of paper. Do *not* write in this book.
1. A device used to transfer electric power from one circuit to another is called a _____.
2. The transformer has two sets of windings known as the _____ and _____ coils.
3. Draw the schematic symbol for a power transformer.
4. Name three losses that occur in a transformer.
5. Eddy-current losses are greatly reduced by using _____ in construction of the core.
6. The input and output voltages of a transformer vary as the _____.
7. A certain transformer has an input voltage of 100 volts. Assuming the primary has 200 turns, compute the secondary for:

    600 volts_____turns

    10 volts_____turns

    20 volts_____turns
8. Explain why alternating current is used in the transmission of electric power.
9. Draw the circuit of an induction coil.
10. Draw the circuit for a door buzzer.

Home appliances, like microwave ovens, use transformers to reduce and increase the 120 volts coming out of wall outlets. Lower voltages are used for the digital display, and higher voltages are needed by the magnetron tube that generates microwaves to cook food and heat liquids. (RCA)

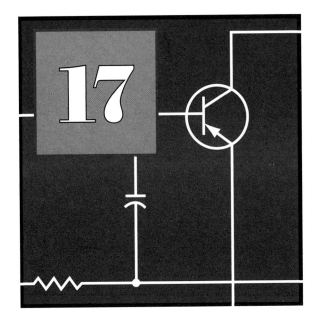

Semiconductors

The transistor was invented in 1948. This tiny electronic component has revolutionized the electronics industry. It has paved the way for unbelievable miniaturization of circuitry and equipment. The transistor forms the basis of integrated circuits that have helped to place a human being on the moon and have provided us tremendous computing power at our desktops. This chapter will explain the background of the transistor, and all that it has made possible.

Objectives

After studying this chapter, you will be able to answer these questions:
1. What is a solid-state device?
2. What is a solid-state diode?
3. How does a transistor amplify a signal?

Important Words and Terms

The following words and terms are key concepts in this chapter. Look for them as you read this chapter.

alpha (α)
amplification
base
beta (β)
bridge rectifier
collector
cutoff
diode
doping
driving signal
emitter
filter
forward bias
full-wave rectifier
half-wave rectifier
hole
NPN
N-type material
PNP
P-type material
reverse bias
saturation
transistor
valance electrons

Amplification

In 1906 a new invention appeared on the American scene. It was the *audion*, invented by a famous American scientist, Dr. Lee DeForest. This was a three-element vacuum tube, which could perform as an amplifier, a detector of radio waves, and a generator of radio frequency waves. This date marks the birth of radio technology.

The ability of a component or device to increase the magnitude of a very small audio or radio signal wave is called *amplification*. The ability of a vacuum tube, and more recently the transistor, to

History Hit!

Lee DeForest (1873–1961)

A graduate of Yale University, DeForest learned and wrote about radio waves. His efforts at the Western Electric Company centered on thermionic (vacuum) tubes. He took a two-element tube called a vacuum tube diode and inserted a third element into it. This created a tube referred to as an audion, or vacuum tube triode. This device, when properly biased (the proper voltages and polarities applied to it), could be used to amplify tiny signals. From that time until the invention of the transistor in 1948, the triode was the primary method to perform amplification in electronic circuits.

amplify is an outstanding contribution to electronics technology and is directly responsible for the spectacular growth of the industry.

Figure 17-1 illustrates the meaning of amplification. A small signal (ac sine wave) is introduced to the input of the amplifier. The output wave from the amplifier is increased in size or amplitude. A signal can be amplified several times until the amplitude is satisfactory for a particular application.

It is important to understand that the added power of the output wave of an amplifier comes from a power source connected to the amplifier. The input signal is the *driving signal,* which causes the amplifier to produce the output signal.

Before we study the transistor and its ability to amplify, an introduction to semiconductor, or "solid-state," devices should first be accomplished.

Semiconductor Current

In earlier studies, we discovered that an electric current could be explained as the transfer of energy along a conductor by *electron* movement. This statement must be revised to explain conduction in some types of semiconductors. In some materials, current carriers will be holes. *Holes* have a net positive charge and attract negatively charged electrons. The diagram in **Figure 17-2** explains electron movement. **Figure 17-3** explains hole movement. Current in a transistor can be either electron or hole flow, depending upon the type of material used.

The materials used for semiconductors are *silicon* and *germanium.* These are both semiconductor, or solid-state, materials that are crystalline in nature and they both have four valance electrons. *Valance electrons* are found in the outer most ring, or shell, of the atom. This is the ring with the highest energy level. In order to change their

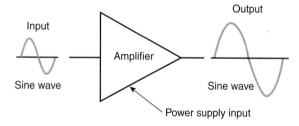

Figure 17-1.
An amplifier increases the amplitude of a signal. The triangle is the accepted symbol for an amplifier. All amplifiers must have a power supply.

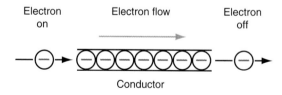

Figure 17-2.
As an electron is forced in one end of a conductor, an electron is forced off the opposite end. Energy is transferred by electrons.

Chapter 17   Semiconductors   **141**

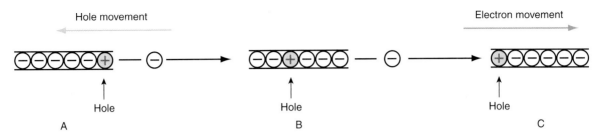

Figure 17-3.
An electron leaving the conductor produces a hole (A) that is filled by the next electron. The hole moves along the conductor (B) until it is filled by an electron from the source (C).

conduction properties from semiconductors, silicon and germanium can be doped. *Doping* supplies a minute quantity of an impurity to change its conduction properties.

N-Type Material
If pure silicon or pure germanium is mixed or doped with a *pentavalent* (five valance electrons) impurity, the number of free electrons is increased. Arsenic and antimony are pentavalent dopants. It then becomes a ***N-type material*** and conduction through it is increased by those negatively charged electrons, **Figure 17-4.**

P-Type Material
If pure silicon or germanium is doped with a *trivalent* (three valance electrons) impurity, the number of holes is increased. Remember, holes are a positive site that can attract the negatively charged electron. Trivalent dopants are aluminum, gallium, and indium. The material then becomes ***P-type material*** and conduction through it is increased by hole flow, **Figure 17-5.**

Diodes

When an N- and a P-type material are joined or grown together, a ***diode*** is created. Diodes can be used to block current in one direction and pass current in the other direction. The symbol for a diode is shown in **Figure 17-6**.

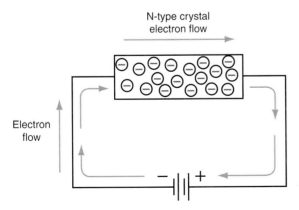

Figure 17-4.
Current flows through the N-type material by movement of electrons.

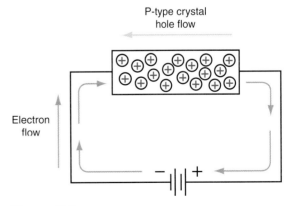

Figure 17-5.
Current flows through the P-type material by movement of holes. Current in external circuit is electron flow.

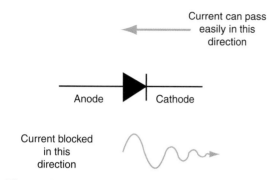

Figure 17-6.
Diode symbol and operation.

Semiconductor Forward Bias

In **Figure 17-7,** the battery is connected so the *negative battery terminal* is connected to the N-type material (Cathode) and the *positive battery terminal* is connected to the P-type material (Anode). This is *forward bias.* The diode acts as a closed switch, and current will flow in the circuit. Electrons leave the source to flow to the N-type material, and conduction through the N-type material is by electrons. At the junction, electrons diffuse across the junction to fill the holes in the P-type material. At the positive terminal, electrons are drawn from the P-type material and are attracted to the positive battery terminal. When an electron leaves the P-type material, a hole is created.

Holes drift toward the junction where they are filled with electrons.

Semiconductor Reverse Bias

In **Figure 17-8,** the voltage is reversed. The *negative battery terminal* is connected to the P-type material and the *positive battery terminal* is connected to the N-type material. This condition is known as *reverse bias.* The diode acts as an open switch. Electrons in the N-type material are attracted toward the positive battery potential. Holes in the P-type material are attracted toward the negative battery potential. As a result, electrons and holes do not have an opportunity to meet and recombine (switch places) at the junction. Very little current (almost zero) flows in the circuit. *A diode is a one-direction only conductor.* It is represented by schematic symbols in both Figures 17-7 and 17-8.

Light-Emitting Diodes

The *light-emitting diode (LED)* is a special diode that gives off light when the correct voltage is applied to the P-N junction. The schematic symbol and bias connections are shown in **Figure 17-9.**

As electrons travel in forward bias through the P-N junction in a LED, energy is given off in the form of light. Light-emitting diodes can produce light at

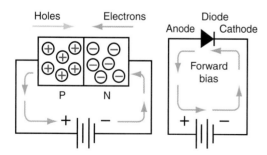

Figure 17-7.
The diode in a forward bias connection. The P-type material carries current by holes. The N-type material carries current by electrons.

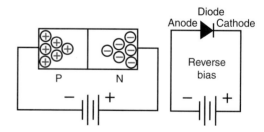

Figure 17-8.
There is almost no current in the reverse bias connection. Holes and electrons do not meet at the junction to recombine and provide current.

Chapter 17 Semiconductors

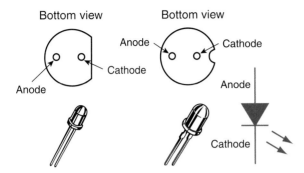

Figure 17-9.
Bias connections and schematic symbol for the light-emitting diode (LED).

various wavelengths producing light that is red, green, yellow, and infrared, **Figure 17-10**. They are very inexpensive, easy to manufacture, and have extremely long life. Another advantage is that they can be turned on very quickly (a few billionths of a second) as compared to a filament lamp. They operate from very low voltage and currents in the circuit.

The light-emitting diode can become more than just an *on* indicator. For example, bidirectional diodes can emit light of two different colors. The bidirectional LED is really made up of two different diodes made with different dopants in the semiconductor material. If the current is in one direction, one of the LEDs emits light energy, red perhaps. If the current reverses, the other LED becomes forward biased and emits another wavelength and color of light

energy, perhaps green. In this way, what appears to be a single LED can tell the status of current or the activity of a circuit.

The seven-segment LED has been used extensively in all digital electronic circuits that must interface with an operator, **Figure 17-11.** Many radios, CD players, video cassette recorders (VCRs), and digital video disc (DVD) players use LEDs made into seven-segment displays to depict the numbers 0 through 9. In fact, these displays, with a few added segments can be made to display alphabetic characters and various symbols. So, a digital circuit can tell the operator, in displayed letters, the status of a task or possible problems within a device. Words like "PLAY" and "RECORD" are shown on a VCR or DVD player.

On many new devices, the LED seven-segment display has been replaced with a liquid crystal display (LCD). This display device uses available light to reflect or not reflect light. Therefore, it uses much less energy than a comparable LED display. Typically, a black segment is observed on a milky white reflective background. This white reflective material is the liquid crystal. By means of an electrical field applied near each segment of the display, light is either reflected or not reflected. This is how the black segments are made to indicate numbers or letters.

Figure 17-11.
Typical seven-segment LED package. Note the inclusion of decimal points before and after the digit.

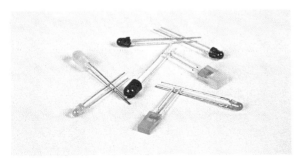

Figure 17-10.
Selection of LEDs of various colors and shapes.

144 Electricity

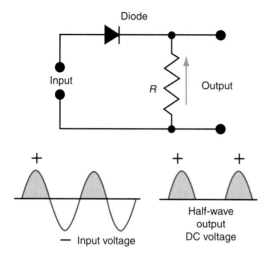

Figure 17-12.
A diode converts an ac voltage to a pulsating dc voltage. The diode conducts only during the positive cycle of the ac input wave. The output waveform represents a half-wave, pulsating direct current.

Rectification

A diode is a very useful component when it is necessary to conve rt an alternating current into a pulsating direct current.

Half-Wave Rectifier

The simple *half-wave rectifier* circuit is drawn in **Figure 17-12.** When the input voltage is on its positive half cycle, the diode conducts (the diode is placed in forward bias), current flows, and a voltage appears across the load resistor. During the negative half cycle, the diode is reverse biased and does not conduct. There is no current in the circuit and no voltage drop across the resistor. The pulsating voltage across the load R is illustrated in Figure 17-12.

Full-Wave Rectifier

Since only half of the wave is used in a half-wave rectifier circuit, more efficient power supplies have been developed to use both halves of the sine wave. These circuits are called *full-wave rectifiers.* In **Figure 17-13**, a full-wave rectifier power supply shows the path of current in the secondary of the transformer from A to B.

Note that diode D_1 in Figure 17-13 is forward biased (conducting) during the positive alternation of the sine wave. Diode D_2 is in reverse bias (nonconducting), and acts as an open circuit. Current is generated in the top half of the secondary coil of T_1 (points A and B) and flows to the load resistor R_1 and back toward D_1. The resistor R_1 sees the current flow and a voltage drop is generated across it.

During the next half cycle, the current reverses direction in the primary and secondary of T_1. See **Figure 17-14.** Now the current flows from C to B in the secondary. At

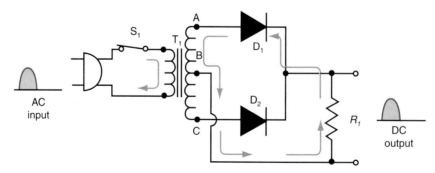

Figure 17-13.
Full-wave rectifier current for the positive alternation of the sine wave.

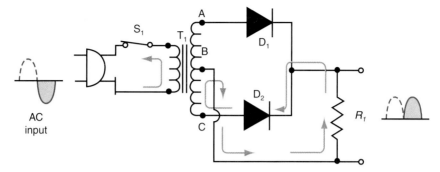

Figure 17-14.
Full-wave rectifier current for the negative alternation of the sine wave.

this time, diode D_1 is in reverse bias (open circuit) and diode D_2 is conducting (closed circuit). Current flows from point C to point B, through the load resistor, R_1, and back to diode D_2. Note, however, that the current still flows in the same upward direction through resistor R_1. Again, as current flows within resistor R_1, a voltage drop is generated across it. The applied ac sine waveform sent to the rectifier from the secondary of the transformer has been changed into full-wave pulsating direct current.

Bridge Rectifier

The *bridge rectifier* circuit is another type of full-wave rectifier circuit that uses four diodes. The primary difference of this circuit and the last is the transformer. The two-diode full-wave rectifier uses a center-tapped transformer for the secondary winding. The bridge rectifier transformer does not need the center tap. It instead uses two more diodes. Oftentimes, these four diodes are packaged within a single case for ease of application.

In the bridge rectifier circuit, two diodes are forward biased at any given time. Likewise, the other two diodes are in reverse bias. **Figure 17-15** depicts a bridge rectifier circuit. Depending upon the polarity of the current flow from the secondary of the transformer, two diodes are conducting, while two diodes are nonconducting. If the polarity at point A is positive, and point B is negative, diodes D_2 and D_4 are conducting. Current is flowing in an upward direction in the load resistor, R_1. Diodes D_1 and D_3 are reverse biased and acting as an open switch. When the polarities change, making point A negative and point B positive, the diodes that conduct and do not conduct switch. In this case, diodes D_1 and D_3 conduct, while diodes D_2 and D_4 are not conducting. Again, the current flows upward through the load resistor, and full-wave pulsating direct current is found at the output.

Filters

Filters often consist of capacitors and inductors. When connected to the output of a rectifier, these devices remove the pulsations in the dc waveform and result in

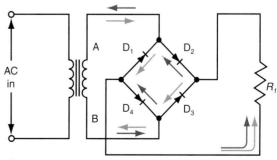

Figure 17-15.
A bridge rectifier circuit with a pulsating dc output.

a constant dc voltage suitable to operate most electronic circuits. A complete power supply circuit is presented in **Figure 17-16.** It uses a transformer to step down the ac input voltage. Two diodes are used in the circuit for full-wave rectification. A resistor and capacitor are used for the filter. R_2 is called a bleeder resistor. A *bleeder resistor* will "bleed" off any stored charge placed on a capacitor. Note, if there was no resistor (R_2) in the circuit shown in Figure 17-16, the capacitor would not be able to discharge. The bleeder resistor is a safety device to be sure there is no charge on the output of a power supply after the power supply has been shut off.

Many power supplies today are made with a three terminal voltage regulator. A voltage regulator is actually an integrated circuit made to regulate and help to smooth out the pulses of direct current found in a rectifier circuit. Additional capacitors are used to ensure proper filtering. The voltage regulator can more precisely keep the direct current output voltage constant, regardless of the resistive load attached to it.

Transistors

The bipolar *transistor* is formed with three layers of P- and N-type material in a

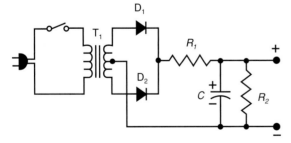

Figure 17-16.
A simple power supply circuit. Notice the transformer, full-wave rectifier, filter (made up of R_1 and C), and bleeder resistor (R_2). Output is a constant dc.

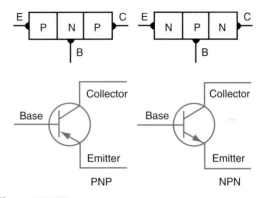

Figure 17-17.
The block diagrams and schematic symbols for the PNP and the NPN transistor. Note collector, base, and emitter designations.

sandwich-like arrangement of two diodes wired back to back. An N-type material between two P-type materials is a *PNP* transistor. If P-type material is between two N-type materials, it is an *NPN* transistor. Both types of transistors are illustrated by block diagrams with their schematic symbols in **Figure 17-17.** Note that three connections are made to the semiconductor material and marked E for *emitter,* B for *base,* and C for *collector.* The base is the center element between the emitter and collector. The transistor is so named because it can *transfer* an electrical signal across a *resistor* or *resistance.*

Transistor Circuits

Three methods can be used in connecting a transistor amplifier. Each method has its own individual characteristics. In **Figure 17-18,** the CB (common base), CE (common emitter), and CC (common collector) circuits are drawn schematically using the NPN transistor. A PNP transistor could be used by reversing the polarities of the batteries or power supplies. In any event, the emitter–base junction is forward biased and the collector–base junction is reverse biased.

History Hit!

Invention of the Transistor

Three individuals are recognized with the invention of the transistor. They are John Bardeen, Walter H. Brattain, and William Shockley. Each man was part of the team from Bell Telephone Laboratories that were co-honored for inventing the transistor. Much theory and experimentation went into the making of the transistor. Of interest is the fact that Bardeen is the only person to receive the Nobel Prize for physics twice. He received the prize once for the transistor and once for shared research into superconductivity. Brattain was born in China and was known to be a practical physicist. Shockley was born in London. After arriving in the United States, he directed research into anti-submarine warfare during World War II. Along with Bardeen and Brattain, Shockley attempted to find a replacement for the energy-hungry vacuum tube. They succeeded. Each shared in the Noble Prize for physics in 1956 for the invention of the transistor.

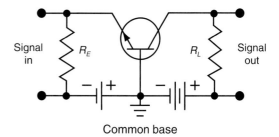

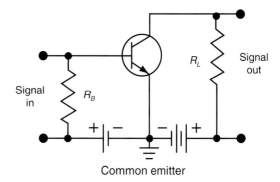

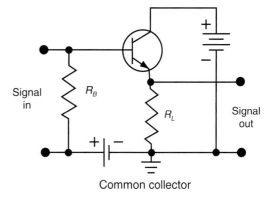

Figure 17-18.
The schematic diagrams for the common base, common emitter, and common collector transistor circuits.

In the *common base* circuit, the input signal is connected between the emitter and *base*; the output is taken from the collector and *base*. The base is common to both input and output circuits.

In the *common emitter* circuit, the input signal is applied to the base and *emitter*; the output is taken from the collector and *emitter*. The emitter is common to both input and output circuits.

In the *common collector* circuit, the signal is applied to the base and *collector*; the output is taken from the emitter and *collector*. The collector is common for both input and output circuits. Note that the collector is not at dc ground. The collector is usually grounded for signal purposes by a capacitor with low reactance connected to the collector. The emitter is isolated from the collector by the load resistor.

Transistor Amplifiers and Alpha

How a transistor amplifies can be understood by examining **Figure 17-19.** The input current is marked I_E. This current is carried through the emitter section by electrons (in the NPN type transistor). At the emitter–base junction electrons combine with the holes in the base section. The emitter–base junction is forward biased, but the base section is very thin and lightly doped. Many more electrons enter the base section than can be combined with the holes. These many electrons feel the influence of the higher positive voltage applied to the collector and flow across the base–collector junction through the collector section and onward to the positive voltage source, V_{CC}. In fact, only about one or two percent of the total current, I_E, flows in the base circuit as I_B. The remaining 98 to 99 percent flows to the collector circuit as I_C. The current I_C produces a voltage drop across R_L (load resistor) that is the output of the amplifier.

The relationship would appear as,

$$I_E = I_B + I_C$$

and I_C must always be less than I_E by a small amount.

Note the terminology used in the transistor. Large amounts of current are emitted from the emitter (E), and collected at the collector (C). The control element is the base (B).

The relationship of I_C/I_E is called *alpha* (α) and means the current gain of the common base transistor circuit. Because collector current is always slightly less than emitter current, the ratio of I_C/I_E will always be less than one. A relatively small voltage is required at the input to increase I_E. Most of this current becomes I_C and produces a larger voltage across the larger resistance R_L. In this manner, a large gain in voltage can be realized.

Common Emitter Circuit and Beta

When the transistor is connected in the common emitter circuit, we find that the alpha current gain is still useful. An example illustrates this point. Assume in **Figure 17-20** that 10 mA is flowing in I_E,

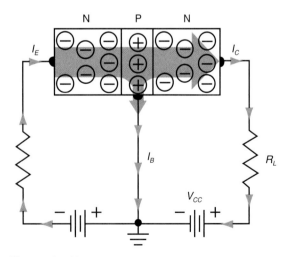

Figure 17-19.
Current through the transistor. Only a small fraction of the current flows in the base circuit.

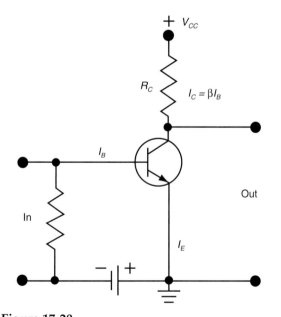

Figure 17-20.
In the common emitter circuit, a small increase in I_B produces a large change in I_C.

two percent or 0.2 mA is flowing as I_B, and the remainder 9.8 mA is I_C. This tells us that there is a relationship between I_C and I_B. This relationship is called *beta* (β):

$$\beta = \frac{I_C}{I_B}$$

or

$$\frac{9.8 \text{ mA}}{0.2 \text{ mA}} = 49$$

Beta is the current gain of the common emitter circuit. Within the operating range of the transistor, if I_B changes a small amount, then I_C will change beta times as much or: $\beta \times I_B = I_C$. In this manner, the common emitter circuit produces a *current gain*. It also produces a *voltage gain* as I_C flows through R_C. Remember only a small voltage at the input will cause an increase in I_B.

This brief description of transistor theory is an introduction to the field of solid-state electronics. It should encourage you to continue your studies and discover the many and varied applications of transistors in electronic equipment.

The Transistor Switch

The transistor can amplify a tiny signal into a much larger signal. It can also operate as a switch. If the base or control element is not providing current, there will be no current flowing in the transistor (from emitter to collector). In this way, an unbiased base-emitter junction would provide the same electrical circuit as an open switch (no current). In a transistor, this operation is called *cutoff*. All current has been cut off from flowing.

Likewise, if the base emitter–junction is made to handle the maximum amount of current possible, maximum current will flow from the emitter to the collector. This is called *saturation*, because the transistor has been saturated with current.

The switching operation was one of the first uses for the transistor. Vacuum tubes used for switches were first used in early computers. Transistors replaced vacuum tubes in use as switches. Now transistors are being replaced by integrated circuits made up of thousands of switches.

Other Special Semiconductor Devices

This section provides introductory information related to some special semiconductor devices. Basic circuit functions are explained along with information related directly to these devices.

Silicon Controlled Rectifier

A *silicon controlled rectifier (SCR)* is a heavy-duty switching device, designed to handle large amounts of current and voltage. Switching devices of this type are known as *thyristors*. As its name states, the SCR is a rectifier. Therefore, this device is commonly known as a dc switch.

Figure 17-21 shows an SCR and its schematic symbol. Note, it has an anode and cathode like any other diode. However,

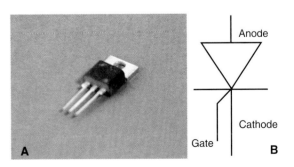

Figure 17-21.
Silicon controlled rectifier. A—Typical SCR component. B—SCR symbol.

Figure 17-23.
Symbol for a diac. The circle may or may not be included with the diac symbol.

it has a third electrode called the gate lead. The gate lead is the control element used to turn the SCR on. A positive pulse on the gate lead activates the SCR. The SCR then conducts until the anode voltage is removed, reduced, or reversed. The SCR is able to *latch-on.*

In other words, the SCR, once turned on, stays on until the voltage on the anode lead is removed, reduced, or reversed. Primary current flow is from the cathode to the anode.

Triac and Diac

A *triac* is known as an ac switch with three terminals (tri-ac). In schematic form, **Figure 17-22,** it looks like two diodes wired in parallel with opposite polarities. It also has a gate lead. The primary path of the current is in the main terminals (MT), referred to as MT_1 or MT_2. Because the triac looks like two diodes wired in opposite directions, once gated or turned on, current can flow in either direction (from MT_1 to MT_2 or from MT_2 to MT_1). The control element is the gate lead. It must receive a pulse on the gate in order to turn on. Because the triac handles ac, it turns off as the sine wave alternates between positive and negative values (when the voltage is 0).

A *diac* is a two terminal device. It is always used in the gate lead of a triac. The purpose of the diac is to provide a triac with a uniform gate pulse. It is this gate pulse that turns on the triac. The diac, when connected to a positive voltage level, drops in resistance value and provides the pulse needed to turn on the triac. See **Figure 17-23.**

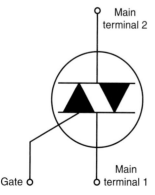

Figure 17-22.
Symbol for a triac. The circle may or may not be included with the triac symbol.

Unijunction Transistor

The *unijunction transistor (UJT)* is used much like a diac. That is, the UJT is used as a trigger device and can activate other circuits. It is made up of a base of N-type

---

Web Wanderings!

http://www1.nasa.gov/home/

At the National Aeronautics and Space Administration (NASA) Web site, you will find the latest news on space exploration and aerospace technologies. Click on "For Students" to view articles on aerospace-related technologies and how those NASA-developed technologies have been put to use on earth.

Chapter 17 Semiconductors

material. However, a small dot of P-type material has been planted on one side. This is the side of the emitter lead. The UJT can vary its resistance depending upon the voltage applied to the emitter lead. In this way, the UJT can act as a triggering device for various circuits.

It can also act as an oscillator. An oscillator is a device that outputs a fixed uniform waveform found most often in timing circuits. **Figure 17-24** depicts the schematic symbol of the UJT.

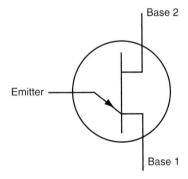

Figure 17-24.
Symbol for a UJT. The circle may or may not be included with the UJT symbol.

Quiz–Chapter 17

Write your answers to these questions on a separate sheet of paper. Do *not* write in this book.

1. The ability of a component or device to increase the magnitude of a very small audio or radio signal wave is called _____.
2. P-type material has a shortage of _____.
3. A device used to conduct electrical current in only one direction is called a(n) _____.
4. LED stands for _____-_____ _____.
5. Give four advantages for a light-emitting diode.
6. A(n) _____ is a useful component when it is necessary to convert a(n) _____ current into a pulsating _____ current.
7. A rectifier can change alternating current into _____ current.
8. The transistor is formed with _____ semiconductor layers in sandwich-like arrangement.
9. Three methods can be used in connecting a transistor amplifier. They are common _____, common _____, and common _____.
10. Which three-terminal device is used as a switch to handle large amount of dc voltage?

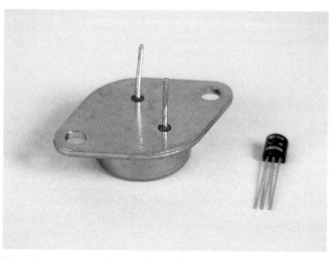

Transistors come in a variety of sizes. The larger transistor can handle greater currents.

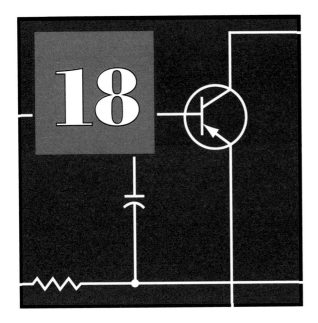

Integrated Circuits

Objectives

After studying this chapter, you will be able to answer these questions:
1. What is an integrated circuit?
2. What is the meaning of a linear or analog circuit?
3. What is meant by a digital circuit?
4. What are computer system input devices?
5. What are computer system output devices?
6. Why is memory so important in relation to a computer system?

Important Words and Terms

The following words and terms are key concepts in this chapter. Look for them as you read this chapter.

analog circuit
AND gate
digital circuit
input device
integrated circuit (IC)
linear integrated circuit
memory
microprocessor
NAND gate
NOR gate
NOT gate
operational amplifier (op-amp)
OR gate
output device
random-access memory (RAM)
read-only memory (ROM)

After the invention of the transistor in 1948, the technology developed to add multiple components such as diodes, transistors, and resistors, to a single material called a substrate. The substrate or base is usually made of the semiconductor silicon. This process was referred to as *integrating* the circuit with many components into one package. Thus, the name of *integrated circuit (IC)* was coined and the microelectronics age began.

Integrated circuits have an advantage over individual discrete components in that many components can be packaged into a very small area. The classification scheme for integrating transistors onto a single chip or substrate from small scale integration to very large scale integration is shown in **Figure 18-1.** In fact, more than 100 million transistors can now be placed into an IC for special applications (such as aerospace applications). This substitution of microelectronic devices for discrete components greatly reduces the cost and size of electronic devices such as cellular phones, CD players, and any device made with ICs.

A key use of integrated circuits is in computer technology. Many powerful devices

Type of Integration	Approximate Number of Transistors per Chip
(SSI) Small Scale Integration	Up to 10
(MSI) Medium Scale Integration	Up to 100
(LSI) Large Scale Integration	10,000–20,000
(VLSI) Very Large Scale Integration	100,000 or More
(ULSI) Ultra Large Scale Integration	Over 1 Million

Figure 18-1.
Classifications of types of integrated circuits.

have been built using the IC. The microprocessor is one such device. ICs can take several forms. As devices become smaller and smaller, so too must ICs be reduced in size. **Figure 18-2** depicts a selection of IC packages used to control a computer's hard drive.

Basing Diagrams for ICs

Figure 18-3 shows a typical basing diagram for two common types of integrated circuits. Note that the index or reference point is the starting location for the numbers, and the pins are numbered in a reverse or counterclockwise direction when viewed from the top surface.

Digital and Linear Integrated Circuits

Integrated circuits can be classified into two major types: digital and linear. A *digital circuit* is either on or off depending on the input. Usually a large number of digital circuits are needed to process information such as in calculations and other computer applications.

Linear integrated circuits, on the other hand, require amplification of the input signal. The output of a linear device is

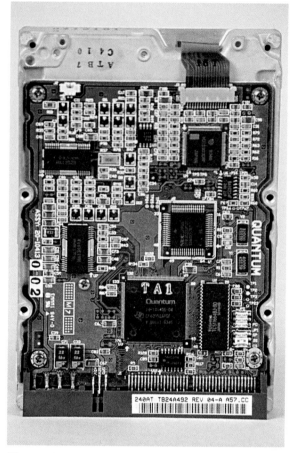

Figure 18-2.
Shown is one side of a computer's hard drive. Notice the different shapes and sizes of the square and rectangular IC packages (in black).

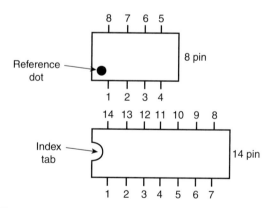

Figure 18-3.
Basing diagrams for IC's. Both are examples of a dual inline package (DIP).

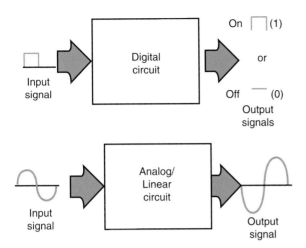

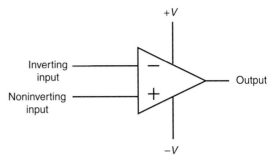

Figure 18-5.
The schematic symbol for the operational amplifier. Note: bipolar power supply leads are not usually depicted in all schematics ($+V$, $-V$).

Figure 18-4.
Digital and analog/linear circuits compared.

variable, depending upon the input. These ICs are not just on or off as in digital ICs. Linear devices can also provide amplification or *gain* to a circuit. Often times, linear circuits are referred to as *analog circuits.* Both terms describe a signal or waveform that rises and falls to many different levels. See **Figure 18-4.**

Linear Circuits

One of the most commonly used linear ICs today is the *operational amplifier,* or *op-amp.* As its name states, it is able to perform operations such as add, subtract, multiply, and divide. One primary purpose is to multiply or amplify a smaller signal. The symbol for an op-amp is found in

Figure 18-5. This device has two inputs and a single output. Integrated into this 8 pin dual inline package or DIP IC is about 22 transistors, a number of resistors, and a capacitor. The op-amp is typically powered by a bipolar power supply. This means a positive and negative voltage is needed to operate the device.

A typical circuit with an op-amp is found in **Figure 18-6.** This circuit shows an inverting amplifier. This circuit amplifies a tiny signal but also changes (inverts) the input wave shape. Note in Figure 18-6 the increase in size of the waveform. In addition, observe a phase change of 180 degrees.

By making a simple change to the circuit, **Figure 18-7,** the op-amp can amplify a tiny input signal without inverting it.

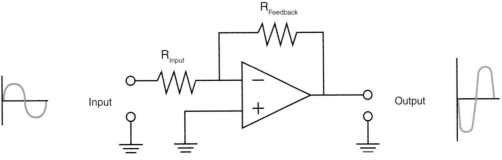

Figure 18-6.
The op-amp inverting amplifier.

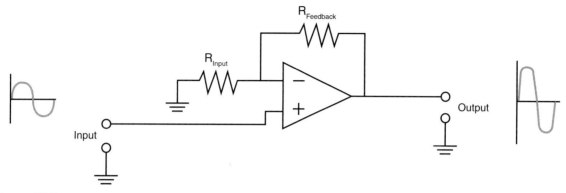

Figure 18-7.
The op-amp noninverting amplifier.

Note that the resistors can be used to change the gain or amplification of the circuit.

These devices all need an external power source, which is usually not shown in the drawings. This makes the schematic diagram a bit easier to read.

Digital Circuits

Digital circuits do not really concern themselves with sine waves as do linear or analog circuits. Instead, digital circuits only work with two circuit values. These values can be called on/off, high/low, or typically, by the logic voltage levels of +5 volts and 0 volts (ground or common). These two values are called binary logic levels and can be combined together to make decisions based upon the inputs.

Basic binary logic gates (gates are circuits that open and close to binary values) can often be represented by simple electrical circuits. Such a circuit is shown in **Figure 18-8.** Here, an *OR gate* is depicted. Note that the lamp can be turned on by either switch A or switch B. The function of the OR gate is just like the switches in the circuit. If input A *or* B has a high voltage value on it, the output of the OR gate is also high (typically +5 volts). If both inputs are a low or 0 volt level, the output is low. If both inputs are high, the output is again high. Hence, the OR gate is operating like switch A and switch B. The output is decided upon by the inputs. Note that there is usually only one unique output state.

The *AND gate* is a bit different from the OR gate. Notice **Figure 18-9.** Here, both switch A *and* switch B must be closed to turn on the lamp. The AND gate, in order to have a high level on its output must have two high levels on the inputs. A low on either or both inputs results in a low output from the gate.

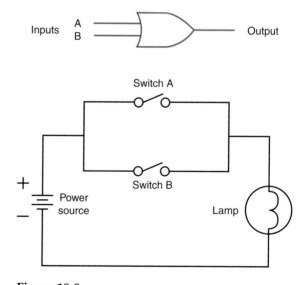

Figure 18-8.
Top—Symbol of an OR gate. Bottom—Switching circuit that compares to the OR gate function. The lamp is on when either switch A OR B is closed.

Chapter 18 Integrated Circuits

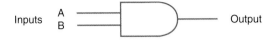

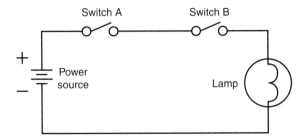

Figure 18-9.
Top—Symbol of an AND gate. Bottom—Switching circuit that compares to the AND gate function. Both switches A AND B must be closed for the lamp to light.

the AND gate. The output is a high logic level whenever either input is a low logic level. The only way to get a low logic level from a NAND gate is to place a high on both inputs.

Notice that all of these digital logic gates mentioned so far have only two inputs. Digital logic gates of these types can have two, three, or more inputs. However, they will all behave, or combine input values in the same way. All of these types of gates will only have one output.

The last logic gate to consider is the *NOT gate*, **Figure 18-11**. This device has only one input and only one output. Its purpose is to invert, which means to take the opposite of the input. A high logic level input becomes a low logic level output. A low logic level input becomes a high logic level output.

From these simple logic gates, decisions can be made that form the basis of all calculations and operations of a digital or computer circuit. For example, many safety interlocks on equipment involve digital decision making. A robotic application that involves the stacking of boxes must not injure workers who might be in the area. Should an employee cross the path of a moving robot, he or she could be seriously hurt. Sensors, that provide a high or low logic level, could sense the presence of the worker. Should the worker cross the potential robot pathway, the sensor would change its output value and, via digital information, shut down the robot.

Enhanced digital gates are the *NOR gate* and the *NAND gate*, **Figure 18-10**. These devices function in the NOT mode. Whatever the function of a gate is, take the opposite of it or NOT it. So, in order to get the high logic level from the NOR gate, both inputs must be low. If any input is a high logic level, the NOR gate's output is low. This is just the opposite of the OR gate.

The same holds true for the NAND gate. A NAND gate is just the opposite of

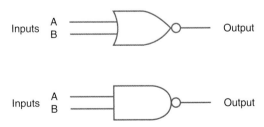

Figure 18-10.
Top—NOR gate. Bottom—NAND gate. Note the addition of a small bubble on the output of each gate. This causes the NOT function of each gate.

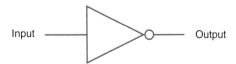

Figure 18-11.
The NOT, or inverter, gate. This device takes the opposite of any digital input value. Note that there is only one input and one output.

Computers

Computers are devices that can save, process, and retrieve data. Modern computers come in many different sizes and with a variety of abilities. Small computers are often used to control appliances. Large computers track and predict the weather. Most computers have similar basic structures. The remainder of this chapter looks at these structures.

Microprocessors

The *microprocessor* is the heart of every computer system. What is inside the microprocessor? All the digital logic gates you just learned about, plus a few more. The microprocessor can also be called the CPU or the central processing unit. This is the device that controls the entire functioning of a computer system. This functioning includes a number of additional devices that are connected to the CPU. These additional devices include: input devices, control devices (such as the internal clock or timing circuits and the reset circuit), memory (including read-only memory (ROM) and random-access memory (RAM) sometimes called read/write memory), and output devices. **Figure 18-12** depicts the basic architecture of any computer system.

Input Devices

While the CPU or microprocessor controls a computer system, *input devices*, **Figure 18-13,** tell the microprocessor what to do. Typical input devices are the keyboard and mouse. However, there are a number of other input devices that include optical devices and voice recognition. These input devices form the commands that tell the microprocessor what to do and when to do it. Of course, there is stored information inside integrated circuits that tells the microprocessor what to do once it is turned on or powered up.

Control Devices

The microprocessor must be carefully controlled. If it is not, errors result. Especially important are the timing sequences during operation. Inside the computer system is a circuit called the master clock, or sometimes just called the clock. The clock helps to provide the information related to the time and date as listed on the computer monitor. The master clock is used to sequence or time the various operations of the computer. How quickly numbers are added, functions are executed, and so on, are directly related to the clock. The clock rate can also represent the *speed* of a computer system, or how fast calculations or operations can be made. A fast computer in the past can be painfully slow with the latest software and related applications. This internal clock may run at rates in excess of 2 GHz (2,000,000,000 clock cycles per second). The clock perfectly times the proper merging of information being held as logic levels. Every action and operation of the microprocessor must be properly sequenced and timed or an error will result.

An additional control device is the reset circuit. Here, the microprocessor IC, as well as all other ICs must be reset to particular values. These values are generally the start

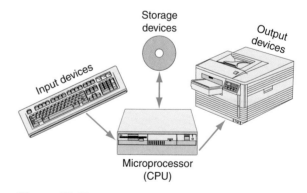

Figure 18-12.
A microprocessor system. Note the flow of data follows the arrows.

Figure 18-13.
A keyboard is one of the most commonly used input devices.

up or wakeup (from being powered off) values. A reset function has significant impact on any operation taking place at the moment the reset button is depressed.

Memory

Memory is typically formed by ICs and other external devices. **Figure 18-14** shows memory media and devices. These devices can be disk drives and on larger computer systems magnetic tape drives. Disks have become smaller and smaller, while the amount of information they can store have become greater and greater.

Read-only memory (ROM) is memory stored on ICs within the computer system. This memory cannot be changed. The information (logic level values) has been imprinted onto the ICs and it cannot be changed. Additionally, ROM does not need to be powered up all the time to keep that information stored. This is called nonvolatile memory storage. What is stored in ROM? Information like what to do upon start up (being turned on). Where do I look for instructions for the procedure to output to the printer? And so on. This basic information is usually referred to as the *boot strap memory*.

Random-access memory (RAM), also called read/write memory, is the memory

Figure 18-14.
Computer memory. A—Information can be stored on discs or in integrated circuits. B—Portable external hard drives can be used to transport large amounts of information between two computers.

Web Wanderings!

http://www.isa.org/

ISA is a nonprofit organization dedicated to providing technical information to manufacturers and employees in the field of instrumentation, systems, and automation. Information on the many educational and training programs provided for their members can be accessed on their Web site. There is also a section for students and educators with links to accredited educational institutions, career information, and a job search database.

where information is stored on a temporary basis only, such as during regular operation of the microprocessor. Once power is removed, this IC memory loses any information stored in it. RAM is called a volatile memory. Once power is lost, so is the data.

Mass data storage devices include various types of magnetic media. This includes magnetic tape, which is usually used with vast amounts of data and in a backup capacity. Magnetic disks have become smaller over the years. At the same time, they are able to store more data.

Today, larger storage hard drives are typically built into the microprocessor system itself. This storage device uses magnetic storage platters to save information. The disk rotates at a much faster speed than other disks, therefore, you are able to access information at a faster rate. There are also external hard drives, see Figure 18-14.

Optical storage devices have become very popular. One of the most common optical devices is the compact disk (CD). It is possible to read and write large amounts of information on compact disks. Optical devices can store more data and there is less chance for damage to the storage medium than with the older magnetic media.

Output Devices

Output devices include monitors, printers and connections made to the Internet or World Wide Web (WWW). Outputs can also be optical and aural (sound). Compact disks are being used more often as both output and input devices. Just as many memory devices can save information, this information can also be read back into the microprocessor by the same device. Both input and output devices are made to be easy to use for the operator of the computer system.

Quiz–Unit 18

Write your answers to these questions on a separate sheet of paper. Do *not* write in this book.

1. Integrated circuits or ICs can be classified into two types: _____ and _____.
2. ICs are made on a base of silicon usually called the _____.
3. Digital ICs are concerned with two logic values, usually referred to as _____ and _____.
4. A commonly used linear IC that provides for amplification is the _____ amplifier.
5. A digital AND gate can only turn on a circuit if both inputs are _____.
6. The NOR gate operates like the OR gate except _____.
7. The NOT gate performs what function?
8. Inside every computer system is a(n) _____ that tells the system what to do and when to do it.
9. The microprocessor can also be called the _____ _____ _____.
10. RAM stands for _____-_____ _____.
11. Nonvolatile memory will _____ after power is removed.
12. List three different output devices.

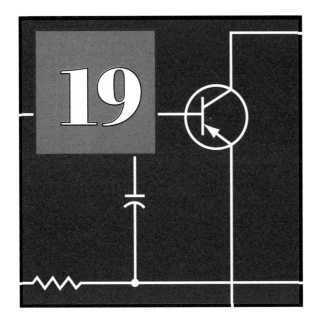

Electrical/ Electronic Projects

Construction Practice

The suggested projects described on the following pages are intended to demonstrate some of the electrical principles you have studied in the chapters of this book.

Before you begin a project, turn to page 7 and review all of the safety procedures. Also, read the soldering primer that is included in this chapter.

The projects are *not* arranged in order of difficulty, and it is *not* necessary to follow a prescribed order.

Exact specifications in respect to size and kinds of material used in these projects have been purposely omitted, thus leaving it to the judgment and desire of the student to adapt and use materials readily obtainable. For example, a buzzer might be constructed on a wood or plastic base. The base might be 3 × 4 or 4 × 6 inches. The exact size is unimportant.

Whatever materials are selected, you should make every effort to develop skills in the use of hand tools so that your projects are well-made. A little care in sanding and finishing a wood base will help produce a project that is worthwhile.

Experience suggests a few ideas that will be useful to you in your activities.

1. Before starting a project, organize your planning notes and sketch your project designs. Review your plans with your instructor, who may give you some helpful suggestions to avoid later difficulties.
2. *Remember: Safety First!* When you are unsure about any electrical circuit, always consult your instructor before testing or energizing *any* circuit.
3. When making a core for an electromagnet, it is important to remember that a low carbon or mild steel should be used. This type of steel does not retain its magnetism. Nails and stove bolts of various sizes make excellent core materials.
4. Wrap the iron core with a layer of electrical tape before you start to wind the coil. This insulates the windings from the core and also provides a good base on which to wind the coil.
5. Insulated wire (enamel covered) should be used for coil winding.

6. Remember that when soldering the ends of enamel-covered wire, the enamel must first be removed. Scrape it off with a knife or remove it with fine abrasive paper. Always wear safety glasses when soldering.
7. In some projects, it is necessary to solder a small piece of iron to a brass or copper armature. This is best accomplished by using a method called *sweat soldering*. Do not use a soldering gun for this operation. It does not get hot enough. To sweat solder two pieces together, first coat each piece with a thin layer of solder. This is called *tinning*. Place the two pieces together in the desired position and apply pressure by means of a short piece of metal rod. While holding the pieces together, apply heat by means of a soldering torch or a soldering iron. When the solder melts between the two pieces, remove the heat but hold firmly in place until the solder solidifies. Observe method in **Figure 19-1**. Always wash your hands thoroughly after handling lead solder.
8. When winding coils, always keep the wire tight and wind it close together. This produces even coils and layers. A suggested method of starting and ending coils is shown in **Figure 19-2**. A small hole is first drilled in the fiber washer near the core. The starting end is put through the hole and extended enough to provide a connection to the coil. When the coil is wound, the finished end is placed through a second small hole in the washer coil end. This holds the coil firmly in place.
9. The springs used in the projects can be made by winding the correct size of music wire or spring brass wire tightly around a form of the correct size. You will need to experiment to get the right spring tension. Hardware stores can also supply various types and sizes of springs.
10. When making a solenoid coil, it should be wound on a hollow core of some nonmagnetic material such as copper, brass, aluminum, plastic, wood, or cardboard.
11. Finished coils can be varnished or lacquered to hold the wires in place and give an attractive appearance. Coils can also be wrapped with electrical tape.
12. Exposed wires in projects can be covered with heat shrink tubing. This plastic tubing is slipped over the exposed wire and/or connection. Then,

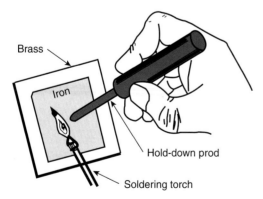

Figure 19-1.
While sweat soldering, pressure is applied to the iron plate by the hold-down prod while heat is applied.

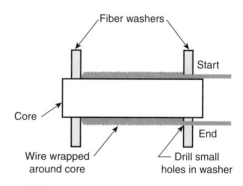

Figure 19-2.
The ends of the coil are inserted in small holes in the fiber washer.

Chapter 19 Electrical/Electronic Projects

> **Safety Suggestion!**
>
> To protect users from electrical shock, it is strongly recommended that all line voltage connections be properly and safely sealed in heat shrink tubing. Be sure to have your instructor review all line voltage connections before power is applied.

heat is applied from a special heat shrink tubing heat gun. The heat provided is enough to shrink the tubing over the exposed connection. Be careful—the heat gun can burn exposed skin and melt plastic and rubber insulation.

13. Metal brackets and coil supports are made easily from soft metals such as aluminum, copper, and brass. Use iron when specified in the plans.
14. Your completed project should include:
    a. Sketch of project with schematic.
    b. A parts list.
    c. Actual working project.
    d. Completed answers to questions on "why it works."
    e. List of several industrial or commercial applications of the electrical principles demonstrated by your project.
15. Always use a low wattage soldering-iron (25–40 watts), when soldering transistors and diodes.
16. Be sure to use a heat sink on temperature sensitive component leads, such as transistors and diodes.
17. It is always a good idea to use sockets for integrated circuits.

Soldering Primer

Knowing how to solder is a must for electrical and electronics technicians. The skill of desoldering is also valuable because components or parts may need to be changed or replaced. For electrical applications, rosin core solder is typically used. This means a fluxing agent (rosin) has been embedded in the solder. This flux aids in cleaning the area to be soldered. Never use acid core solder in electrical work. The acid will eventually destroy thin wires.

Most solder is made up of a mixture of tin and lead. Usually 63% tin and 37% lead is best for electrical or electronic work. Because lead is a dangerous material that is hazardous to humans and to the environment, solder now is made from other alloys (mixtures of different materials) such as a mixture of silver, copper, antimony, and tin. These lead-free solders are comparable to tin/lead solder. To ensure a healthy environment, lead-free solder will soon be used in all electrical/electronic devices.

Before soldering, make sure all wires, connections, and parts are clean. You can use abrasive paper (sandpaper) to clean these components. A good mechanical (physical) connection between wires, circuit board, and components should be made first. Wires should be wrapped tightly around components or other wires. Stranded wires should be twisted and tinned (coated with solder) to prevent the strands of wire from separating and conducting poorly or becoming a hazard.

> **Safety Suggestion!**
>
> Lead is a toxic material. Do not eat, drink, or put hands or fingers in your mouth or near your eyes after handling solder. *Always* wash your hands first. There are special soaps to assist in the removal of lead from your hands after soldering. These cleaning agents can be obtained at specialty stores that provide materials for stained glass window projects.

When inserting components into printed circuit boards, bend the component's leads outward to keep the leads in contact with the circuit board pads (hole with copper track around it) and to keep the component from falling out of the board. This ensures a well-soldered connection.

Once the connection is secured, the following steps are used. It is best to attempt these steps in less than five seconds. This will prevent the connection and electronic part from getting too hot. Some components, such as diodes, light emitting diodes, transistors, and other semiconductors, are more sensitive to heat than others. Five seconds of heat may damage these components. To prevent damage, a heat sink clip should be attached close to the body of the electronic part. The heat sink draws the heat into the clip's metal and away from the component's semiconductor material.

Be sure to read all safety precautions before handling solder. Remember to wear eye protection when soldering. During the soldering process, a smoke or fume will be given off. It is best not to breathe these fumes. *Always* solder in a well-ventilated area.

1. Clean the hot iron tip with a wet sponge or paper towel to remove oxidation as shown in **Figure 19-3**. A clean iron tip will transfer heat more effectively. Be careful not to burn yourself!
2. Melt a tiny amount of solder on the hot iron tip, **Figure 19-4**. This is called tinning the iron and will help to transfer heat.
3. Preheat the connection by tilting the iron tip and placing it against the widest part of the metal or connection, **Figure 19-5**.
4. Add solder by placing it against the heated connection, as shown in **Figure 19-6**, *not* to the tip of the iron. Let the connection melt the solder. It is important not to add too much solder. You do not want to short out other connections. Practice is needed to ensure enough, but not too much, solder is added.
5. Once enough solder has melted into the connection, remove the solder and then the iron. If the iron tip is removed first, the solder may stick to the connection. By removing the solder first, this problem will not happen. A moment after the solder is removed, remove the iron tip. Do not move the connection until the solder has hardened. When the solder has hardened, you can cut off the lead's excess length.

Desoldering Primer

There are two techniques used to remove unwanted solder from a junction. One technique uses a vacuum device such as the vacuum bulb (a rubber or plastic bulb with a plastic tip), the solder sucker (a vacuum device operated by spring pressure), or the desoldering iron (a soldering iron with a vacuum bulb and hollow tip). The most effective and most costly vacuum device is the vacuum desoldering station (a soldering iron and vacuum tip blended together in

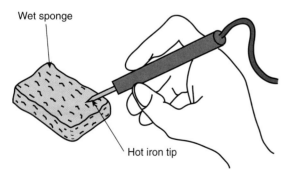

Figure 19-3.
To remove oxidation, hold the soldering iron at an angle and wipe the beveled edges of the hot iron tip across a wet sponge or paper towel. Be sure to get all sides of the tip.

Chapter 19   Electrical/Electronic Projects

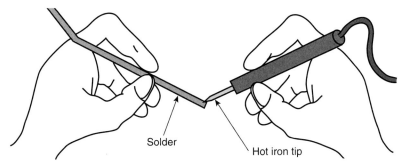

Figure 19-4.
Melt a light coating of solder onto the hot solder tip.

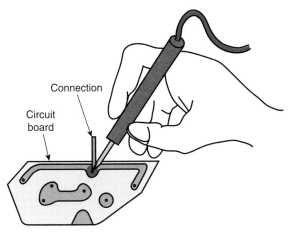

Figure 19-5.
Always preheat the connection before applying solder.

one device). With all vacuum devices, once the solder becomes molten (melted), the vacuum draws the solder into the vacuum device, leaving the junction free of solder.

The other technique uses a copper braid (a mesh of fine copper wires). It is placed between the soldering iron tip and the solder junction. As the soldering iron heats the solder junction, the braid acts as a sponge to absorb the molten solder.

Desoldering with a Vacuum Device

This desoldering primer covers the vacuum bulb, solder sucker, and desoldering iron. If using a vacuum desoldering station, please read and follow the

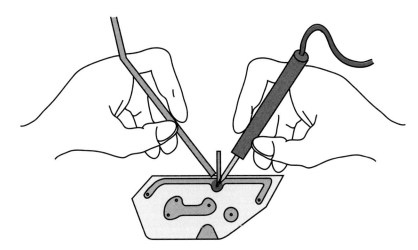

Figure 19-6.
Place the solder against the heated connection. The solder will melt and flow around the base of the connection.

manufacturer's instructions. When desoldering, remember to use a heat sink on heat sensitive components and to complete the operation in five seconds or less.

1. Clean the hot iron tip of the soldering iron or desoldering iron with a wet sponge or paper towel to remove oxidation. Be careful not to burn yourself!
2. Melt a tiny amount of solder on the hot iron tip.
3. Place the hot iron tip against the junction to be desoldered. If using a desoldering iron, depress the bulb before placing the hot iron tip on the junction to be desoldered, not when the iron tip is against the junction and the solder has melted. This will prevent air from expelling from the bulb and blowing molten solder into areas you don't want the solder to flow.
4. Depress the solder bulb or the solder sucker and place it on the solder junction next to the iron tip. If using a desoldering iron, the bulb should be depressed (and held in the depressed position) before applying the iron to the solder junction.
5. When the solder becomes molten, release the suction from the vacuum device.
6. Repeat steps 1 through 5 until all solder has been removed from the solder junction.

Desoldering with a Copper Braid

1. Clean the hot iron tip of the soldering iron with a wet sponge or paper towel to remove oxidation. Be careful not to burn yourself!
2. Melt a tiny amount of solder on the hot iron tip.
3. Place the copper braid over the junction to be desoldered, allowing the end of the braid to extend an inch or two beyond the junction, **Figure 19-7**. Heat will be easily transferred from the copper braid to your fingers—be careful not to burn yourself! Wear gloves.
4. Hold the hot iron tip against the copper braid where the braid touches the solder junction, **Figure 19-8**.
5. When the solder becomes molten, slowly pull the copper braid from between the solder tip and the solder junction, allowing the solder to absorb evenly into the braid, **Figure 19-9**.
6. If there is still solder to be removed from the junction, cut off the used portion of the copper braid and repeat steps 1 through 5.

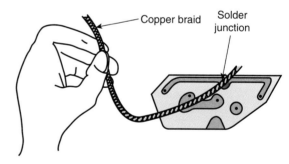

Figure 19-7. An inch or two of copper braid should extend past the junction to be desoldered. This extra length is used to absorb the molten solder.

Chapter 19   Electrical/Electronic Projects   **167**

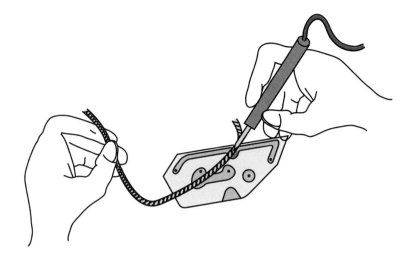

Figure 19-8.
The hot soldering iron tip is placed over the copper braid at the junction to be desoldered. The heat transfers through the braid to the solder junction and melts the solder.

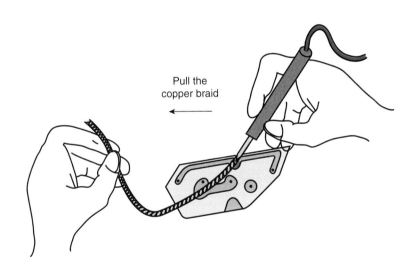

Figure 19-9.
When the solder has melted, pull the copper braid from between the solder joint and the iron tip. Solder will be absorbed along the extra length of the braid.

Project 1
Experimenter

The Experimenter is a valuable electronics tool to use in learning more about how circuits work. See **Figure 19-10**. It has a 5 volt dc power supply, a breadboard or protoboard, and four switches (2 SPST and 2 SPDT). Resistors, incandescent lamps, LEDs, diodes, transistors, capacitors, and other components can be plugged into the socket for tests and experimental circuits. Series, parallel, and series-parallel circuits can also be connected using the Experimenter.

Parts List for Experimenter

SW_1 & SW_2—Single-pole, double-throw (SPDT) rocker switches

SW_3 & SW_4—Single-pole, single-throw (SPST) rocker switches

SW_5—Single-pole, single-throw (SPST) miniature switch

F_1—Fuse holder, panel mounted

Fuse—2 amp, fast acting fuse

NL_1—Neon indicator glow lamp with built-in 100 kΩ resistor, 120 Vac

T_1—Filament transformer: primary 120 Vac, secondary 6.3 Vac @ 1 A

D_1, D_2, D_3, & D_4—Silicon diodes, 1N4001

C_1 & C_2—Electrolytic capacitors, 2200 µF @ 16 Vdc or higher working voltage

IC_1—Voltage regulator (5 V), LM309K (National Semiconductor)

BP_1 & BP_2—Binding posts (one black and one red)

EX_1—Experimenter board

TP_1 through TP_{10}—Tie point block solderless connectors (10)

Chassis—Miniature console style

Misc.—Line cord, perf-board for wiring power supply, solder lugs, rub-on decals, solder

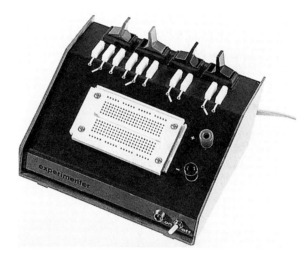

Figure 19-10.
Front and rear view of the Experimenter. Your Experimenter may look different depending on the chassis that you choose.

Chapter 19 Electrical/Electronic Projects

Construction Hints

1. Obtain parts and wire the 5 volt dc power supply. See **Figure 19-11** for the schematic. The line cord should enter the chassis in the back through a rubber grommet. The fuse is mounted in a fuse holder. The toggle switch to turn the power supply on and off is mounted on the front panel.
2. Mount the two single-pole, single-throw (SPST) switches and the two single-pole, double-throw (SPDT) switches on the top of the front panel. Next mount the four tie point block solderless connectors on the chassis (be sure that the pins under the solderless connectors do not short out on the metal chassis). Then wire the switch leads to the four pin solderless connectors.
3. Mount the breadboard on the front panel of the Experimenter.
4. Mount the power supply to the front panel. Be careful that exposed wiring does not touch the metal. Also be sure binding posts do not short out on the metal chassis. Any short circuit will prevent proper operation of the Experimenter and may blow the fuse.

Using the Experimenter

The Experimenter can be used to perform many experiments that can be used to reinforce theory. The following experiments are very basic and will help you to understand the concepts of electricity better. In addition to the experiments in this book, you are encouraged to design your own experiments or to use other experiments that may be provided by your instructor.

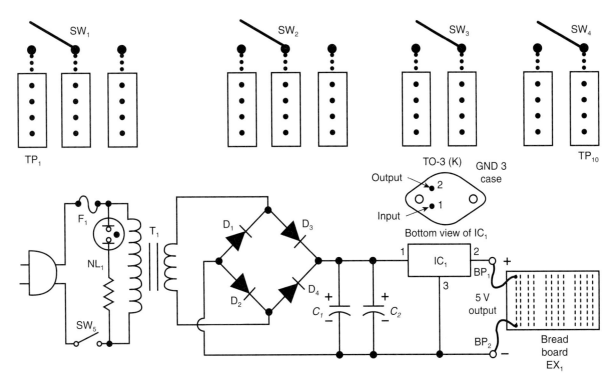

Figure 19-11.
Schematic for the Experimenter.

Safety Suggestion!

Remember, it is much safer to double-check your wiring *before* you turn on the power for an experiment. When using meters, be sure that you have the meter on the correct range and the leads are connected with the right polarity.

Problem 1: Using a Voltmeter and an Ammeter

1. Obtain the following components:
   Experimenter
   Volt-Ohm-Milliammeter (VOM) or Digital Multimeter
   R_1—100 Ω, 1-watt resistor
   Jumper Wires (as needed)
2. Connect the components as indicated in the schematic. (Do not turn on power yet.)
3. Double-check your wiring and then turn on the Experimenter power supply switch SW_5. Next close switch SW_4.

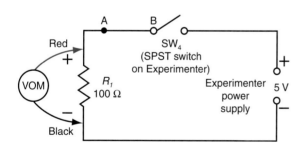

4. As shown in the above schematic, measure the voltage across R_1. The voltage is _____ volts.
5. Next measure the voltage across the output of the power supply (the red and black binding posts). The voltage is _____ volts.
6. Measure the voltage across SW_4. Why isn't there any voltage across SW_4?
7. Insert a dc ammeter in the circuit at points A and B. Be sure to break the circuit so that all the current flows through the ammeter. Start on the highest range and reduce the range setting until you get a reading. The current is _____.
8. Turn off the power and carefully put away the components.

Problem 2: Wiring a Simple Circuit with a Lamp

1. Obtain the following components:
   Experimenter
   Lamp A: #47 lamp (6.3 V @ 0.15 A with 22 gauge solid jumper wire leads soldered to each base connection)
   Jumper wires
2. Connect amp A so that it can be turned on and off by switch SW_4.
3. On a separate sheet of paper, draw a schematic (using the symbols that follow) for a simple series circuit with a lamp.

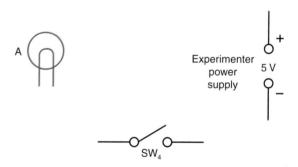

Problem 3: Wiring Lamps in Series

1. Obtain the following components:
   Experimenter
   Lamps A, B, and C: #47 lamps (6.3 V @ 0.15 A with 22 gauge solid jumper wire leads soldered to each base connection)
   Jumper wires

2. Connect lights A, B, and C in *series*, controlled (on and off) by SW_4.
3. On a separate sheet of paper, draw a schematic (using the symbols that follow) for three lamps wired in series.

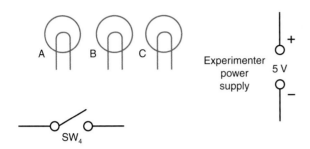

4. Are lamps brighter or dimmer than the one lamp used in *Problem 2*? Explain why.

Problem 4: Wiring Lamps in Parallel

1. Obtain the following components:
   Experimenter
   Lamps A, B, and C (the same as used in *Problem 3*)
   Jumper wires
2. Connect lamps A, B, and C in parallel and controlled by SW_4.
3. On a separate sheet of paper, draw a schematic (using the symbols that follow) for three lamps wired in parallel.

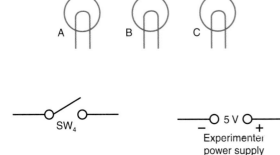

4. Are lamps brighter or dimmer than the lamps in *Problem 4*?
5. Which circuit, *Problem 3* or *4*, is using the larger amount of electrical power?
6. Are electric lights in your home connected in series or parallel?

Problem 5: Two SPST Switch Control Circuits

1. Obtain the following components:
   Experimenter
   Lamps A, B, and C (the same as used in *Problem 3*)
   Jumper wires
2. Connect lights A and B so that SW_4 will control light A and SW_3 will control B.
3. On a separate sheet of paper, draw a schematic (using the symbols that follow) for two SPST switches controlling two lamps independently.

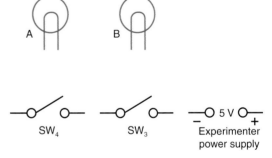

Problem 6: Wiring a Three-Way Switch Circuit

1. Obtain the following components:
   Experimenter
   Lamp A (the same as used in the last problem)
   Jumper wires
2. Connect light A so that it can be turned on and off by either switch SW_1 or SW_2 (three-way switch or SPDT switch).

3. On a separate sheet of paper, draw a schematic (using the symbols that follow) for the control of a single lamp by two switches.
4. Why are three-way switches used?
5. Name some locations in your home in which three-way switches are used or could be used.

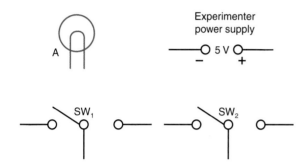

Project 2
Audio Oscillator

The audio oscillator is used to generate a tone. There are many applications for oscillators in electronics circuits. You will find this project in **Figure 19-12** and schematic in **Figure 19-13** both interesting and useful.

You have studied the theory of operation of a transistor. A transistor can conduct a current or resist the flow of current, depending upon conditions under which it is operated. In transistor circuits, a direct current voltage is applied to all three elements: the collector, the base, and the emitter. The current flowing through the base and one of the other elements can be used to control a current flowing through any other two elements. For example, the current from base to emitter can control the current flow from emitter to collector.

Students interested in the hobby of amateur or "ham" radio will find this code practice oscillator practical. Coupled with a telegraph key, the code oscillator is a useful tool while building up Morse code speed. Because this audio tone oscillator is variable

Figure 19-12.
Audio oscillator.

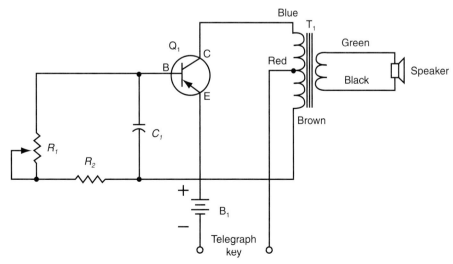

Figure 19-13.
Code oscillator schematic.

in pitch, some students learn to play simple tunes on it.

Parts List for Audio Oscillator

R_1—Potentiometer 25,000 Ω

R_2—Resistor 6800 Ω, ½-watt

C_1—Capacitor 0.5 µF, 200 V, paper

T_1—Audio output transformer—Primary: 500 Ω to 1000 Ω C/T; Secondary: 8 Ω

Q_1—Transistor—PNP general purpose transistor low power (Germanium): 2N567 or SK3004.

B_1—9 volt transistor battery

Permanent magnet speaker—2 inch, 8 Ω

Minibox enclosure—4 3/4″ × 2 1/2″ × 1 2/5″ with cover.

Misc.—Plug/jack for key, control knob, and miscellaneous hardware

Construction Hints

1. The audio oscillator in the photograph in Figure 19-12 was constructed in a plastic minibox. You may select a container of another size or material. Be sure you have enough room to easily mount all the components of your audio oscillator.
2. See **Figure 19-14** for printed circuit layout and transistor basing diagram.
3. The speaker used is one of the 2 inch small transistor permanent magnet speakers.
4. The oscillator is powered with a 9 V transistor battery.
5. Check connections against the schematic.
6. Ask your instructor to check your work.

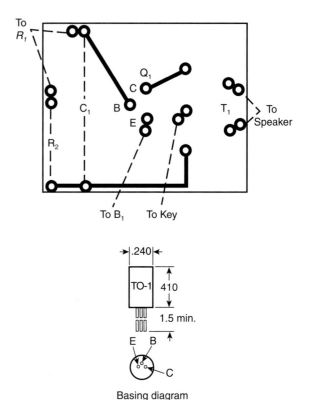

Figure 19-14.
Printed circuit layout for audio oscillator.

Project 3

Transistor Radio Receiver

In Project 2, you used the transistor as an oscillator. To further demonstrate the versatility of the tiny transistor, the simple radio receiver shown (schematic shown in **Figure 19-15**) can be built. In this receiver, the modulated radio signal is detected by a crystal diode and the transistor is used as a stage of audio amplification.

Much of the fascination in building this radio lies in making it as small and compact as possible. It is suitable for local broadcast reception.

You can experiment with the antenna. Many students use a short piece of wire and an alligator clip. The clip is fastened to most any available metal object to act as an antenna. The radio has worked well with the clip attached to a bicycle, a window screen, and so on. A ground connection improves the reception, but it works quite well without one.

Parts List for Transistor Radio Receiver
L_1—Antenna coil with center tap
C_1—Variable capacitor, 365 pF
C_2—Capacitor, disc 0.02 µF (20nF)
C_3—Capacitor, Paper, 0.1 µF (100nF)
R_1—Resistor 470 kΩ, ½-watt
R_2—Resistor 1500 Ω, ½-watt (Note: Use R_2 if crystal earphone is used. Not needed if magnetic earphone is used.)

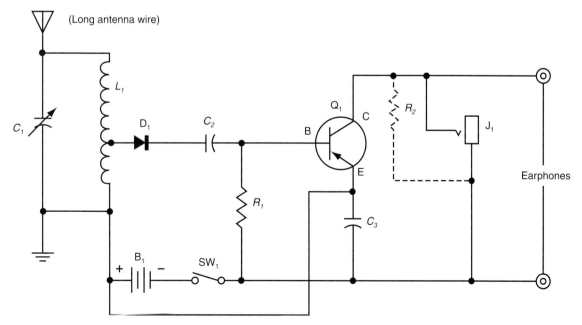

Figure 19-15.
Diode transistor radio receiver.

Crystal earphone—2000 Ω

Phone tip jacks

Switch—SPST, push button or miniature toggle

B_1—1.5 volt AA cells in series (2)

Antenna & ground terminals

D_1 — Diode 1N34

Q_1 — Transistor, PNP general purpose transistor low power (germanium), 2N107 or SK3003.

J_1 — Jack (optional for phone tip jacks)

Misc. —Tuning knob

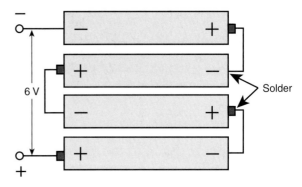

Figure 19-16.
Solder leads on the ends of the positive and negative terminals connect these AA cells in series.

Construction Hints

1. A case for the radio can be purchased at an electronics parts supply store or manufactured out of metal or wood.
2. Most any battery combination up to six volts can be used. Use two or four AA cells connected in series, **Figure 19-16.** Wrap the cells up in electrical tape to make a neat bundle.
3. Wire in the usual manner, tracing your schematic with a red pencil as each connection is made. This will ensure that you did not miss anything.
4. Ask your instructor to inspect your wiring.
5. Connect a long antenna wire to the radio (connection of C_1 and L_1), turn on the switch, and tune C until a station is heard.

Project 4
Door Chime

An application of the solenoid is the familiar door chime. The chime illustrated in **Figure 19-17** is a double chime and does require more mechanical work than a simple single chime.

An added principle of mechanical motion is used in this project. When the coil is energized, the plunger is pulled into the coil and strikes one chime. When the coil is de-energized, the plunger is returned to its original position by the action of the spring. The momentum of the returning plunger causes it to overshoot its at-rest position. This overshooting permits the plunger to strike the second chime.

Construction Hints
1. You will want to do some experimenting with the kinds of tubes or bars you use for your chimes. Generally a thin tube will produce the most pleasant tone. The tubes or bars can be cut most any length. Using a longer length gives a lower or deeper tone. In Figure 19-17, flat pieces of hard steel are used.
2. Holes have been drilled in the chime bars and rubber grommets inserted in the holes. Supports are designed to hold bars in position, but not rigidly. The coil construction is explained in the illustration, **Figure 19-18**. The core is 1/4 in. copper tubing and the ends or coil supports are made of 20 gauge sheet metal or plastic. The core and the end brackets are soldered or glued together.
3. Wind about four layers of enameled insulated #22 wire for the coil.
4. The chime support pin, **Figure 19-19**, is one way of hanging the chime tubes or bars. A fine wire also works well. The tubes should not be supported rigidly as this will deaden the ringing tone. They should hang straight and vertical.

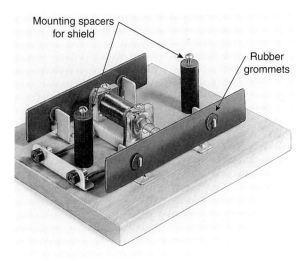

Figure 19-17.
The completed double action door chime.

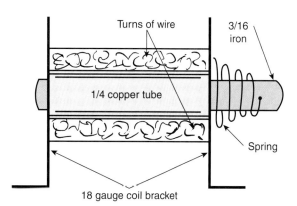

Figure 19-18.
Coil construction of two-tone door chime.

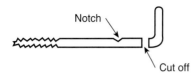

Figure 19-19.
One method of making a pin to support a chime tube or bar.

5. The plunger is made of a common nail and the cone-type spring is wound with #6 music wire (piano string). The spring is fastened to the core by drilling a small hole and inserting one end of the spring into it.

6. Some careful adjustments in spring tension and placement of the chimes are necessary to permit the plunger to strike the chimes on both movements of the plunger. While these adjustments are being made, the chimes should be hung on the wall and in a vertical position. For decoration and appearance, a shield can be made of metal or plastic, to cover the working mechanism of the chimes. The shield can be polished, tooled with a design, or left plain. The shield is mounted with long screws or bolts and held in front of the chime by tubular spacers.

Project 5
Magnetic Relay

One of the more common applications of magnetism is found in the relay. A relay, in its simple form, is nothing more than a switch operated by magnetism. It has some distinct advantages. One of major importance is its ability to turn on or off a high voltage or a high current machine or device from a remote location. This provides a high degree of safety for working personnel. Only a small low voltage current is required to activate the coil. The magnetism of the coil operates the armature that can have large and heavy contact points for switching on and off the high power machine. The entire relay can be enclosed for greater protection. The relay also has the ability of rapid switching when required.

Construction Hints

1. Relays may be constructed in many ways. The photograph of only one kind appears in **Figure 19-20.** Note that it is a single-pole, double-throw switch. The switch contacts are connected to the three binding posts with color-coded wires. The relay can be used as a *normally closed* relay and will open the circuit when the coil is energized or a *normally open* relay that will close the circuit when energized. Which contacts to use will depend on its application. (See **Figure 19-21.**)
2. The coil is wound with three layers of #22 enameled wire around a plastic spool. A 1/4 inch × 2 inch bolt is used as a core.
3. The frame of the relay is made of 16 gauge aluminum. The switch contacts are mounted on plastic for insulation and the plastic is bolted to one side of the aluminum frame.
4. The armature is also plastic and is placed on top of the opposite side of the frame. It is held down by the tension spring. A round piece of iron is cemented to the plastic armature over the core of the coil.
5. Notice the bolts and nuts used for the contact points. Adjust the top contact to change the gap (spacing) between the coil and the armature.
6. The tension spring is also adjustable so that more or less current is required for relay operation.
7. Wires connected to the operating coil are connected to the other two binding posts.

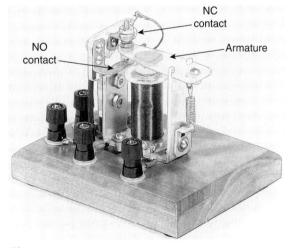

Figure 19-20.
A completed experimental relay.

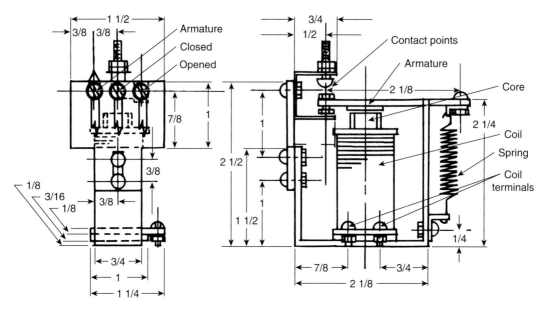

Figure 19-21.
Suggested dimension of magnetic relay.

8. A schematic of the relay in **Figure 19-22**, shows the electrical connections.

Relay Operation

1. Connect the relay coil to a variable (0–6 V) voltage source. Connect a voltmeter across the source and an ammeter in series with the coil as shown in **Figure 19-23**.
2. Connect a 6 V lamp to the normally closed contact point and a 6 V power source. The light should be *on*.
3. Starting with *zero* volts to the relay coil, slowly increase voltage until relay operates. The indicating lamp will go out.
4. Record the voltage and the current required to operate the relay: _____ Volts, _____ mA.
5. Return voltage to zero. Adjust tension spring so that tension is slightly increased. Repeat steps 3 and 4. _____ Volts, _____ mA.
6. Adjust the upper contact point to push the armature downward. This decreases the air gap (spacing) between the coil and the armature. Repeat steps 3 and 4. Record the voltage and current. _____ Volts, _____ mA. What effect does changing the gap have on relay operation? Explain.

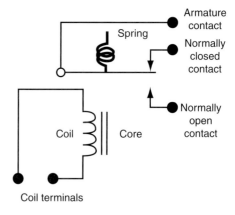

Figure 19-22.
The electrical diagram of connections to parts of the relay.

Chapter 19 Electrical/Electronic Projects

7. The relay can be used to switch any kind of a device or machine on or off. The contact points must be heavy enough to carry the required current of the machine. In **Figure 19-24,** the relay is used to switch a motor.

The Buzzer or Doorbell

With slight changes, the relay can be used as a buzzer. Connect the circuit as drawn in **Figure 19-25.**
1. Connect one end of the relay coil to the armature contact.
2. Connect the source voltage to the other end of the coil and to the *normally closed* switch contact.

When the buzzer or doorbell push button is closed, the relay is energized. This pulls down the armature and *opens the circuit*. This allows contact points to return to the *closed* position and energizes the relay again. This opening and closing of points in rapid succession produces a *buzz*. The frequency of the buzz can be changed by adjusting the gap spacing or the spring tension.

3. Measure and record the voltage required to run your buzzer.

If a striker is attached to the vibrating

> **Safety Suggestion!**
>
> Whenever line voltage is connected to a circuit, be sure your instructor checks all your wiring and connections for proper and safe operation.

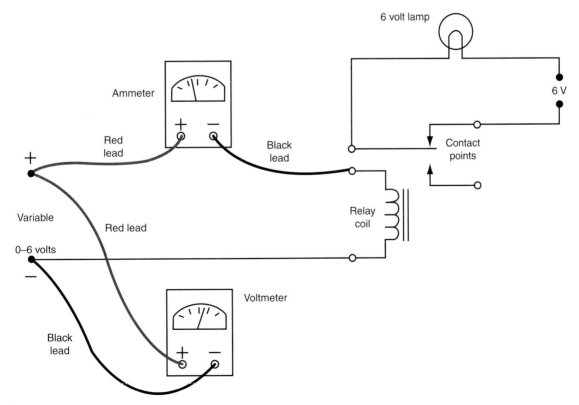

Figure 19-23.
The electrical connections for testing the operating voltage and current of the relay.

182 Electricity

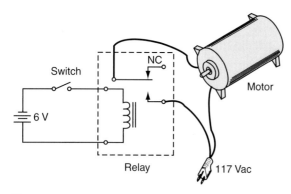

Figure 19-24.
The relay is used to remotely control a motor.
Note: Exposed line voltage connections in this circuit. Be sure to insulate all exposed connections! Have your instructor check your wiring before you energize the circuit!

armature of the buzzer, it can be arranged to strike a bell. Now you have a doorbell. See the dotted lines added to the armature in Figure 19-25.

Mechanical Oscillator

This project is interesting since it takes advantage of the RC time constant of the circuit for operation. When the proper voltage is applied to the circuit, the lamp will flash.

1. Connect the mechanical oscillator circuit as drawn in **Figure 19-26**. You will need the following extra parts.

   6 volt lamp and screw base
   10 µF electrolytic capacitor, 25 volt or higher voltage rating
   SPST switch

2. Connect the mechanical oscillator to a variable power source. Increase the voltage gradually until the lamp starts to flash (about 15 volts). The switch must be closed.

3. Using the circuit diagram, trace the current when relay points are normally closed and open. When the switch is open, a current flows through the closed points and the capacitor charges. This current is too small to light the lamp. When the switch is closed, a voltage is applied to the relay coil that causes the armature to change to the other contact and the lamp will light. Now this opens the coil circuit, and we would expect it to operate like a buzzer. It does not, however, since the capacitor must discharge and this discharge current flowing through the coil delays the action of the relay.

4. Remove the lamp from circuit and replace with a 4 Ω speaker. Now you can hear the oscillations.

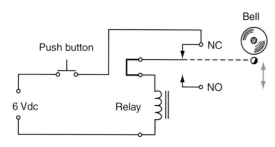

Figure 19-25.
The relay connections to make a buzzer or doorbell.

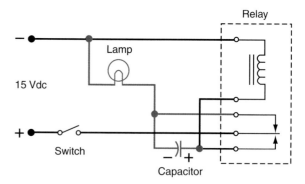

Figure 19-26.
The diagram for a mechanical oscillator using RC time constants.

Chapter 19   Electrical/Electronic Projects

Project 6

Induction Coil

A project that has a good deal of educational value is the induction coil, **Figure 19-27.** Coils of this nature are widely used. One of the more common applications is the ignition, or high voltage, coil in your automobile.

The principle of transformer action is well demonstrated in this project. You should study the chapter on transformers so that you will understand the action of this circuit.

If a varying current is flowing through a coil, an ever-changing magnetic field exists about the coil. The magnetic field rises and falls in step with the magnitude of the alternating current.

If a second coil is placed near the first coil, the first coil's magnetic field will cut across the second coil. This will *induce* a voltage in the second coil. The first coil is

Safety Suggestion!

The surge of voltage from an induction coil can cause a serious electrical shock. Be sure *not* to touch the secondary coil wires when the primary is energized.

called the *primary*. The second coil is called the *secondary*. The voltage induced in the secondary is in direct proportion to the *turns ratio*. For example, ten turns on the primary and one hundred turns on the secondary will produce ten volts across the secondary for every one volt applied to the primary.

In our project, six volts are applied to the primary. The output voltage will depend on the number of turns you use on the secondary coil.

Actually the induction coil, without the secondary, is just another buzzer. The buzzer action is necessary to cause the rise and fall of current through the primary. Without this changing current, there can be no transformer action.

When the induction coil is operating, a small spark can be made to jump across a gap between wires connected to the secondary.

Construction Hints
1. The core is made of 1/16 inch welding rod cut to 6 inch lengths, grouped together and wrapped with electrical tape to form a round core.

Figure 19-27.
Photo of an induction coil.

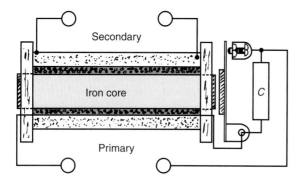

Figure 19-28.
Construction plan for the induction coil.

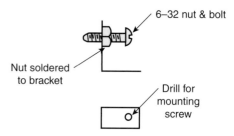

Figure 19-30.
Detailed plan for induction coil contact points.

2. The coil ends can be either wood or plastic. A hole the size of the core is drilled in each end. The ends are cemented in place on the core, **Figure 19-28.**
3. The primary winding consists of two layers of #18 enamel covered wire. Wind neatly and then cover with another layer of electrical tape. *Leave about 6 inches of wire at each end of the coil for later connections.*
4. The secondary is wound with several layers of #26 enamel covered wire. The greater number of turns will produce a higher induced voltage output on the secondary. *Be sure to leave sufficient wire at the end of the coil for later connections.*
5. The construction of the armature and points is illustrated in **Figures 19-29** and **19-30.**
6. Wiring instructions can be found by tracing the schematic drawing in **Figure 19-31.**

A new component has been added to this circuit that requires some explanation. This is the capacitor, C. You should study the chapter on capacitor action in a circuit, so that you understand its purpose. The capacitor serves two useful functions. First, when the points open, we know that an arc will be created across the points. The capacitor will absorb this electrical energy and eliminate the arc across the points. This causes a more rapid change in the magnetic field and consequently more efficient operation of the induction coil. Secondly, the charged capacitor tends to reverse the flow of current through the coil, when the points open. This makes the magnetic field disappear very rapidly and again makes a more effective induction coil. The capacitor used in this project is a 0.1 µF paper capacitor. You may wish to try other values and observe the effects.

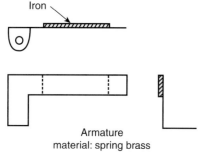

Figure 19-29.
Detailed plan for induction coil armature.

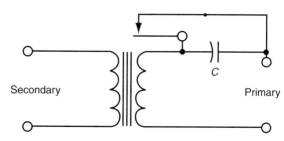

Figure 19-31.
Schematic diagram of the induction coil circuit.

Project 7
Two-Pole Motor

There are many types and kinds of motors. Some operate on direct current, while others use alternating current. The motor converts electrical energy into mechanical energy. The extensive use of motors at home and in industry suggests that an elementary understanding of such a device is essential to our education in electricity.

The theory and operation of the motor is explained in Chapter 11. As you construct this small motor, an awareness of the underlying principles of operation will be helpful.

A motor can be built in many ways. If you wish to build one like the one in **Figure 19-32**, follow the plans carefully. It would be better if you adapted the materials at hand and constructed the motor on your understanding of the principles.

Construction Hints

1. The motor illustrated in Figure 19-32 was built on a wooden base about 4 inches × 5 inches, but any size base of a suitable material will be satisfactory. Sand and finish your base.
2. The armature can be made first. Refer to drawing in **Figure 19-33**. Secure a 5/16 inch hexagon nut. Lay out, center punch, and drill in two opposite flat sides of the nut a hole the exact size of a 20d nail. Cut two 20d nails to length of about 1 1/4 inch. Place two 3/4 inch fiber washers over each nail and force the nails into the holes drilled in the nut. If the nails fit too loosely, place the end on an anvil and hit it sharply with a hammer. This will knock the nail slightly out-of-round and cause the nail to fit tightly in the holes.
3. Spread the two fiber washers on the nails and place a layer of electrical tape over the nail between the washers. Starting at the center by the nut, wind the individual coils. Wind about six layers of #22 enamel covered wire on

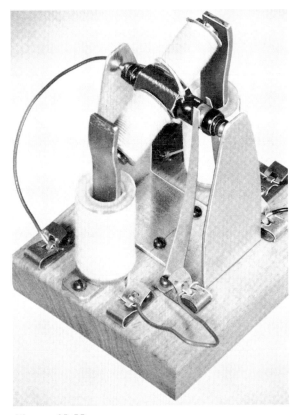

Figure 19-32.
A two-pole ac motor that can be connected as a series or shunt wound motor.

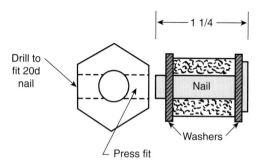

Figure 19-33.
The nails are forced into the holes in the nut to form the cores of the armature.

each armature coil. Both coils should be wound in the same direction. They are connected together in series. It is not necessary to cut the wire between the coils. Wind one coil and bring the wire across the nut to the start of the second coil and continue winding.

4. The armature shaft is made by cutting the heads from a 5/16 inch × 1 1/4 inch and a 5/16 inch × 3/4 inch bolt. Place the bolts in a lathe or drill press and file the unthreaded ends to a cone-shaped point. These points will be the motor bearings. One bolt is screwed halfway into the nut from one side and the other bolt is screwed into the other side. Jam the two bolts together by tightening with pliers or a pipe wrench.

5. The threaded portion of each bolt is now wrapped with a layer of electrical tape for insulation and appearance. The commutator sections are made by sawing a piece of 1/4 inch copper tubing lengthwise. The two copper pieces are placed over the longer side of the armature shaft. If the fit is not tight, build up the shaft with another layer of tape. Neatly solder each of the two wires coming from the armature coils to a commutator section. Secure the sections in place by wrapping the ends with tape. Be sure that a space remains between the sections. About one quarter inch should be left untapped in the center so that brushes can make contact with the commutator. See **Figure 19-34**. *Important: The placement of the commutator sections in relation to the coils must be correct.* See **Figure 19-35**.

6. The motor frame is made with 16 gauge aluminum bent in a U-shape, **Figure 19-36**. The armature is placed in position between the uprights of the bracket. Lay the bracket on its side on the bench and hit it with a hammer. The cone-shaped points will make

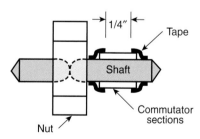

Figure 19-34.
The two commutator sections are fastened to the shaft with plastic tape.

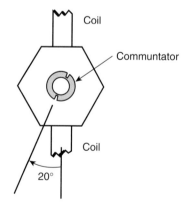

Figure 19-35.
Commutator sections are turned to an angle of about 20 degrees from center line through coils.

Chapter 19   Electrical/Electronic Projects

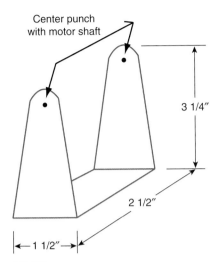

Figure 19-36.
The motor frame.

indentations in the aluminum and form the bearing holes for the revolving armature. Fasten the bracket and armature assembly to your base with round head wood screws.

7. The field poles of the motor are made by shaping two pieces of 1/8 inch × 5/8 inch × 4 1/4 inch iron. Three-quarters of an inch from one end, bend the iron to a right angle. Drill a hole and mount the angles to the wooden base. Your armature should revolve freely between these poles and in line with them yet not touch them. Some shaping and bending may be necessary. A slight curvature to conform to the revolving armature will make your motor more powerful.

8. The field coils are wound on spools 1 inch diameter by 1 3/4 inch long. A 5/8 inch hole is drilled through the center of the spools so they slide over the field poles. The coils are wound with six layers of #22 enamel covered wire. The ends of each coil are connected to Fahnestock clips fastened to the base on each side of the coil, **Figure 19-37**.

9. The brushes are made of 22 gauge brass (other materials and thicknesses can be used) 3 1/2 inch long and about 1/8 inch wide. Bend a right angle on one end of each brush, drill and fasten to base with round head wood screw. A Fahnestock clip is also fastened to each brush with the same mounting screw. The brushes should line up and make contact with the commutator sections. Slight bending adjustments may be necessary.

10. Your motor is now ready for a trial operation. The individual Fahnestock clips for each field coil and the brushes allow you to connect the motor as a series or shunt wound motor. They also provide a method of reversing the motor rotation. Connect a jumper wire between the two field coils so that the outside winding on one coil is connected to the outside winding of the second coil. Connect the remaining terminal of each field coil to the brush terminals by jumper wires. Attach a 6 V source across the brush terminals. It may be necessary to turn the armature slightly to start its rotation. Your motor should run if you have followed the instructions carefully.

Two-Pole Motor Experiments

Refer to the diagram in **Figure 19-38** for connections and lead identifications for these experiments.

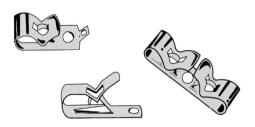

Figure 19-37.
Fahnestock clips. (General Cement Mfg. Corp.)

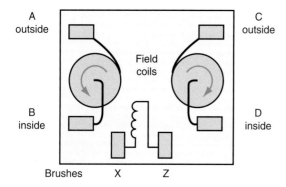

Figure 19-38.
Two-pole motor connections.

Experiment 1. Connect as series motor: C to A, B to X. Apply voltage to D and Z.
  A. The motor runs clockwise or counterclockwise?
  B. Connect wires on schematic diagram for the series wound motor.

Experiment 2. Connect as shunt motor: X to B, A to C, D to Z. Apply voltage across brushes X and Z.
  A. The motor runs clockwise or counterclockwise?
  B. Draw another schematic diagram like that of *Experiment 1* showing the connections for the shunt motor.
  C. Reverse the connections of battery on terminals X and Z. The motor runs clockwise or counterclockwise?
  D. When connections are made as in question C, the field pole with coil AB is a north or south pole?
  The field pole with coil CD is a north or south pole? (Hint: Use left-hand rule.)

Experiment 3. Reverse polarity of fields in the shunt motor. Connect A to X, C to Z, and A to D. Apply voltage across X and Z. Does the motor run clockwise or counterclockwise?

Experiment 4. Connect the motor in the following manner: C to B, X to A, Z to D. The motor runs clockwise, counterclockwise, does not run? Explain.

Project 8
Electric Engine

The easy-to-build engine shown in **Figure 19-39** demonstrates the conversion of electrical energy into reciprocating motion and then to rotary motion by means of a connecting rod and flywheel. The engine operates by the solenoid action of the coil, which is energized when the breaker points close. The breaker cam on the end of the flywheel shaft should be made adjustable so that the engine can be timed. For proper operation, the points should close early, on the inward stroke of the solenoid piston. The points must open just before the maximum inward stroke. The kinetic energy stored in the flywheel carries the engine through its no-power stroke. **Figure 19-40** shows the timing for proper operation.

Figure 19-39.
An electric engine that demonstrates solenoid action and engine timing.

Construction Hints
1. Dimensions and exact construction are unimportant in this project. Think the problem through and design your own base and mechanical parts.
2. The coil is wound with six layers of #20 enamel covered wire over a 3/8 inch copper tubing core.

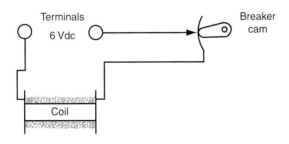
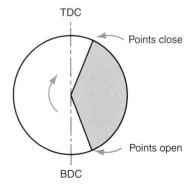

Figure 19-40.
Left. Connection diagram for electric engine opening and closing in relation to top dead center.
Right. A timing diagram showing top dead center (TDC), bottom dead center (BDC), and timing points.

3. In Figure 19-39, the coil ends have been shaped from small blocks of wood.
4. For the flywheel, a section 1/4 to 1/2 inch long can be cut from a 1 1/2 to 2 inch round steel bar. A lathe can be used for facing the flywheel.
5. Shaft supports can be bar iron or aluminum.
6. The cam is fitted snugly to the shaft, but you should be able to rotate it for timing of the engine.
7. The breaker points are constructed of 24 gauge brass.
8. Connections are diagrammed in Figure 19-40.

Project 9
Continuity Tester

The simple continuity tester project shown in **Figure 19-41** will help you determine opens or shorts in electrical circuits. It can be used to test low voltage circuits with power. Also, it is self-powered and can be used to test continuity.

Parts List for Continuity Tester
R_1—470 ohm, 1/4 watt resistor

LED—Light-emitting diode (Maximum forward current: 30 mA; maximum forward voltage 2.5 V)
D_1—Diode, Silicon Switching Diode Fast Recovery (t_{rr} = 4 ns), 1N4148 or SK3100
B_1—9 V transistor battery
SW_1—SPDT switch
Misc.—Plastic box with cover, wire, test leads with probes or alligator clips on one end and banana plugs on other end

Construction Hints

Refer to the parts list and **Figure 19-42** of the schematic for the continuity tester. Note that the battery is used in one position of switch SW_1. In the other position of SW_1, the continuity tester requires the power of the circuit to be tested. Observe the correct polarity of the light-emitting diode when you wire the circuit.

To operate the continuity tester, place

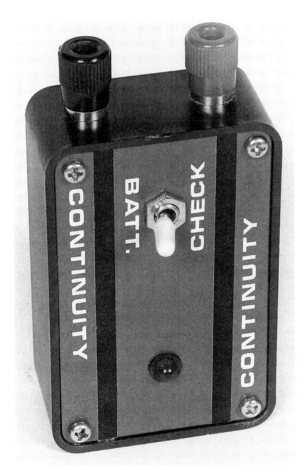

Figure 19-41.
Continuity tester.

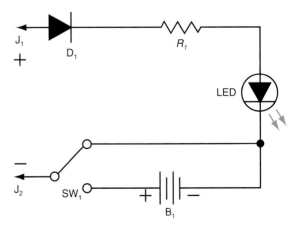

Figure 19-42.
Schematic for continuity tester.

the switch in the "battery" position. Touch the two leads together and the LED should glow. This indicates a complete circuit in this mode. It can be used on disconnected or unpowered circiuts across switch terminals to test whether switches are on or off or faulty. Also, fuses can be tested to see whether they are open or not. Suspected broken wires in a circuit can also be checked for continuity with this testing device.

In the "check" position (with the tester's battery disconnected), the circuit relies on external voltage for operation. In this mode, the continuity tester can be used to test batteries from about 4.5 volts to 15 volts. Be sure to connect the probes in the circuit with the correct polarity. It can also be used to test whether low voltage dc self-powered circuits are operating properly.

Project 10

Automotive Battery Charger

Everyone that owns a car should have a battery charger. This inexpensive and useful project can charge a 12-volt automobile battery overnight, **Figure 19-43**. It produces 1.5 amps of current and acts as a "trickle," or slow, battery charger. Refer to the parts list and **Figure 19-44** of the schematic for the automotive battery charger. Like the Experimenter, this project operates off of line voltage and current. Be sure to take all safety precautions. *Make sure all exposed high voltage connections are covered with heat shrink tubing or electrical tape. Have your instructor check all wiring.*

Parts List for Automotive Battery Charger

SW_1—SPST switch

NL_1—110 Vac neon pilot assembly (with built-in resistor)

T_1—Transformer, power, 110 Vac primary; 12.6 Vac @ 3 A secondary, center tapped

D_1 & D_2—Silicon rectifiers, 3 A @ 50 PIV, 1N5400 or SK9003

F_1—Panel mount fuse holder and fuse (3 A)

Clips #1 & #2—Heavy-duty battery clips

Case—Plastic with aluminum cover

Misc.—Line cord and plug; heavy-duty cable for clips #1 & #2; grommets; decals

Construction Hints

The automotive battery charger is very easy to use. Simply connect the positive clip of the charger to the positive lead of a car battery and the negative clip of the charger to the negative lead. It is recommended that the charger be connected to the battery for a minimum of 6 to 8 hours in order for the trickle charge to be effective.

As with any lead acid storage battery, there should never be an open flame in the vicinity. Explosive hydrogen gas is produced during the charging process and it can be easily ignited by an open flame. Additionally, the battery fluid is an acid that must be handled safely. Be sure not to touch your face or eyes if you contact the fluid. Immediately wash your hands when you are done. The corrosion that can form on a battery terminal can also be acidic. Take special precautions such as wearing

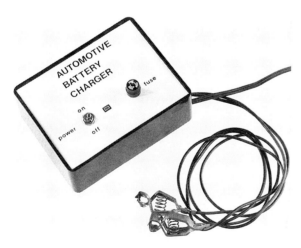

Figure 19-43.
Automotive battery charger.

194 Electricity

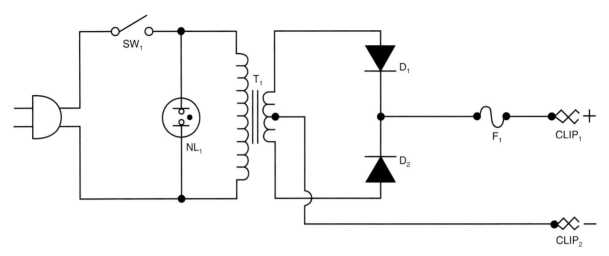

Figure 19-44.
Schematic for automotive battery charger.

gloves and eye protection when working with automobile batteries.

Do not try to jump-start an automobile with this charger since such a low output current (1.5 amp) is provided. Additionally, it is best to have the battery completely removed from its circuit (disconnected) when it is being charged. This automotive battery charger is an excellent slow charging device for 12 volt rechargeable batteries.

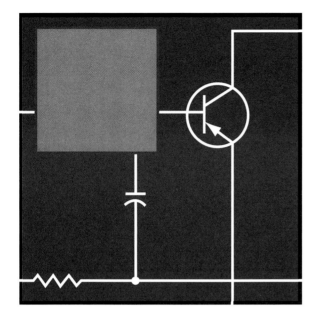

Glossary

A

alpha (α): Current gain of a common base transistor circuit. Collector current divided by emitter current, I_C/I_E.

alternating current (ac): Current of electrons that moves first in one direction and then in the opposite direction.

ammeter: Meter used to measure current. Connected in series in the circuit.

ampere (A): Unit of electrical current. One ampere equals one coulomb (6.24×10^{18} electrons) passing a given point in a conductor in one second.

ampere-hour (Ah): Unit of measurement of the capacity of a cell or battery. A battery with one hundred ampere-hours capacity can theoretically supply one hundred amperes of current for one hour, or one ampere for one hundred hours.

ampere-turn (At): Unit of measurement of magnetomotive force. Represents the product of amperes times the number of turns in a coil of an electromagnet.

amplification: Ability of a component or device to increase the magnitude of a small ac electrical signal.

analog circuit: Circuit whose output varies in proportion to the input, unlike a digital circuit that has only two logic states. Also referred to as a linear circuit.

AND gate: Type of digital circuit that must have a high logic level at both inputs to provide a high logic level output.

armature: Movable part of a generator or motor, usually the rotating windings and magnets. In a relay or buzzer, the armature is the movable arm in the magnetic field.

atom: Smallest particle of element that can exist without losing its identity.

atomic number: Number of protons in the nucleus of an atom.

atomic weight: Mass of nucleus of an atom in reference to oxygen, which has a weight of 16.

autotransformer: Transformer with a common primary and secondary winding.

average value: Value of alternating current or voltage of a sine waveform. Found by dividing the area under one alternation by the distance along the X-axis between 0° and 180°. $E_{avg} = 0.637 \times E_{peak}$.

B

base (B): Thin section between emitter and collector of a transistor.

battery: Group of two or more cells connected together. A single cell is sometimes called a battery.

beta (β): Current gain of the common emitter transistor circuit. Collector current divided by base current, I_C/I_B.

bridge rectifier: Type of rectifier (a circuit that converts ac into dc) that uses four diodes and a non-center tapped transformer to change an ac input signal into a full-wave pulsating dc output.

brush: Sliding contact, usually of copper and carbon, that rubs on the commutator or slip ring of a generator or motor. Through brushes, an electrical connection is made to a revolving armature.

C

capacitance (C): Inherent property of electric circuit that opposes change in voltage.

capacitive reactance (X_C): Opposition to alternating current as result of capacitance.

capacitor: Device that possesses capacitance. A simple capacitor consists of two metal plates separated by an insulator.

cell: Single unit consisting of a receptacle containing a negative electrode, a positive electrode, and an electrolyte. The cell generates a voltage by chemical action.

charge: Process of building a negative potential on one plate of a capacitor and a positive potential on the other plate.

circuit breaker: Automatic safety device, usually thermally or magnetically operated, that automatically opens an overloaded circuit.

collector (C): Semiconductor section of a transistor that collects the majority carriers.

commutator: Group of bands providing connections between armature coils and brushes used for reversing direction of an electric current.

compound winding: Method of winding a motor or generator in which the windings are in series and in parallel, either interacting against each other or reinforcing each other.

conductance (G): Ability of a wire or component to transmit or conduct electric current. It is measured in siemens. (Letter symbol for conductance is G.)

conductor: Material that readily transfers electrical energy. It has a very low resistance to electric current.

copper losses: Heat losses in motors, generators, and transformers as a result of wire resistance.

coulomb: Quantity of electrons (6.24×10^{18} or 6,240,000,000,000,000,000 electrons).

counterelectromotive force (cemf): Voltage induced in a conductor moving through a magnetic field that opposes the source voltage.

coupling: Percentage of mutual inductance between the primary and secondary coils of a transformer.

current (I): Transfer or movement of electrical energy in a conductor by means of electrons moving constantly and changing positions in a vibrating manner.

cutoff: When the base of a transistor is not providing current and no current is flowing from the emitter to the collector. A transistor in a cutoff state acts as an open switch.

cycle: One complete reversal of alternating current from positive to negative and back to starting point.

D

d'Arsonval meter: Analog/stationary-magnet moving coil meter.

deflection: Movement of a meter's indicating pointer. The limit of a scale on a meter would be at full-scale deflection.

dielectric: Nonconducting material, such as an insulator.

digital circuit: Circuit whose output varies in only two logic states. These logic states are usually referred to as logic level high (binary 1, +5 volts) and logic level low (binary 0, 0 volts).

diode: Two-terminal device that conducts electricity more easily in one direction than in the other direction.

direct current (dc): Current of electrons that flows in one direction only.

doping: Adding an impurity to a semiconductor material.

driving signal: Input signal to an amplifier. The signal that is to be amplified.

E

E: Symbol for voltage in Ohm's law.

eddy current losses: Induced currents flowing in a body such as a core or armature by variation of magnetic flux.

effective value: Value of alternating current of a sine waveform. It has the equivalent heating effect of direct current. Sometimes referred to as the root-mean-square (rms) value. $E_{eff} = 0.707 \times E_{peak}$.

electric power: Rate of doing electrical work.

electricity: Movement of electrons in a conductor.

electrode: Elements in a cell.

electrolyte: Acid solution in a cell.

electromagnet: Magnet formed by winding several turns of wire around an iron core. The polarity of the electromagnet depends on the direction of the current in the coil.

electromotive force (emf): Force that causes free electrons to move in a conductor. Unit of measurement is the volt.

electron: Negatively charged particle in orbit around an atom.

electroscope: Laboratory device used to detect a small electrical charge. Consists of two small metal foil strips suspended in a glass bottle that repel if charged.

electrostatic field: Static field of force existing around a charged body or terminal.

emitter (E): Semiconductor section in either P- or N-type transistor that emits a majority of carriers into the base.

energy: Capacity for doing work.

equivalent resistance: Expresses the resistance of a network of resistors. Electrically speaking, equivalent resistance is one resistor that equals the sum of several resistors in the circuit.

F

farad (F): Unit of measurement of electrical capacitance. Named in honor of Michael Faraday.

field magnets: Electromagnets that produce the magnetic field of a motor or generator.

field poles: Iron magnets in the field circuit of a motor or generator around which the field windings are wound.

filter: Circuit that changes pulsating dc to nearly pure dc in a rectifier.

flux: Magnetic lines of force running between the poles of a magnet.

forward bias: External voltage applied in conducting direction of a PN junction. The positive terminal is connected to the P-type region and the negative terminal to the N-type region.

frequency (f): Number of complete cycles per second, measured in hertz.

full-wave rectifier: Circuit that uses two diodes to convert alternating current (ac) into full-wave pulsating direct current (dc). It is more effective than the half-wave rectifier because it uses both the positive and negative alternations of the sine waveform.

fuse: Safety protective device made of metal with a low melting point that will melt and open the circuit if current exceeds the rating of the fuse.

G

generator: Device that converts mechanical energy into electrical energy.

H

half-wave rectifier: Circuit that is used to change alternating current (ac) to direct current (dc). It uses only one diode and creates only a half-wave pulsation in the output of the circuit. It is less effective than the full-wave rectifier; however, it is still useful in battery charger circuits.

henry (H): Unit of measurement for inductance. Named in honor of Joseph Henry.

hertz (Hz): Unit of frequency measuring number of cycles per second of an ac waveform. Named in honor of Heinrich R. Hertz.

hole: Positive charge. Space left by a removed electron.

horsepower (hp): Unit of measurement of work accomplished when 33,000 pounds are moved a distance of one foot in one minute, or 550 foot-pounds per second. Electrically, one horsepower equals 746 watts.

hydrometer: Bulb-type instrument used to measure specific gravity of a liquid and determine the quality of a wet cell's electrolyte.

hysteresis: Retardation or resistance to the changing magnetic field in the core material of a generator, motor, or transformer. Molecular friction.

I

***I*:** Symbol for current in Ohm's law.

impedance (*Z*): Total resistance to flow of ac current as a result of resistance and reactance.

inductance (*L*): Property of a coil that offers resistance to a varying current.

inductor: Coil; component with the properties of inductance.

input device: The part of a computer system that is used to enter commands or communications into the computer. Some examples of input devices are a keyboard, mouse, or joystick.

insulator: Material or substance that has a high resistance to the flow of electric current.

integrated circuit (IC): Fabrication process in which many electronic circuits and devices are formed on a single silicon chip.

ion: Atom that has either lost or gained electrons and becomes either positively or negatively charged.

ionization: Process of breaking up atoms into ions. An atom is ionized when it has lost or gained one or more electrons.

IR drop: Voltage loss across resistive components in a circuit. Voltage drop equals current times resistance: $E = I \times R$.

iron vane meter: Meter based on the principle of repulsion between two connective vanes placed inside a coil.

J

joule: Unit of measurement of electrical work or energy. Energy used when one ampere of current flows through one ohm of resistance for one second.

K

kilo (k): Prefix meaning one thousand.

L

laminations: Thin sheets of metal used in cores of transformers, motors, and generators to reduce eddy currents.

law of charges: Like charges repel; unlike charges attract.

***LC* time constant:** Time period required for a voltage across an inductor to increase to 63.2% of maximum value or to decrease to 36.7% of maximum value.

Lenz's law: Induced emf in a circuit is always in such a direction as to oppose the effect that produces it.

light-emitting diode (LED): Solid-state device that emits light if current is passed through the PN junction.

linear integrated circuit: Circuit whose output varies in proportion to the input, unlike a digital circuit. Sometimes referred to as an analog integrated circuit.

local action: Defect in voltaic cells caused by impurities in the zinc, such as carbon, iron, and lead.

M

magnetic field: Invisible lines along which a magnetic force acts. The lines externally leave from the north pole and enter the south pole, forming closed loops.

magnetism: Invisible force of a magnet that causes it to attract iron objects.

magnetite: Magnetic iron ore.

magnetomotive force: Magnetic pressure as a result of the number of turns in the magnet's coil and the current flowing in the coil. Measured in gilberts.

mega: Prefix meaning one million.

memory: Computer hardware device formed by a number of devices (integrated circuits and other external devices), which store information for the computer system.

meter: Instrument used to measure electrical quantities.

microprocessor: The microprocessor is the heart of a computer system. It contains digital circuits that control every operation of the computer.

molecule: Smallest part of a compound, made by the combination of two or more atoms.

motor: Device that converts electrical energy to rotating mechanical energy.

multimeter: Combination voltmeter, ammeter, and ohmmeter. A switch changes the leads to the desired meter and range.

mutual inductance: When two coils are located so the magnetic flux of one coil can link with turns of the other coil. The change in flux of one coil will cause an emf in the other coil.

N

NAND gate: Type of digital circuit whose output is a high logic level whenever there is a low logic level at any input. A high logic level at both inputs, outputs a low logic level. The NAND gate is the opposite of the AND gate.

neutron: Uncharged particle of an atom found in the nucleus.

NOR gate: The output of the NOR gate is a low logic level whenever there is a high logic level at any input. The NOR gate is the opposite of the OR gate.

north pole: Concentration of magnetic lines of force at one end of a magnet. Opposite end and polarity from the south pole.

NOT gate: The NOT gate, which has only one input lead, ouputs the inverted (opposite) input logic level.

NPN: A type of transistor that has a P-type material sandwiched between two N-type material layers.

N-type material: A semiconductor material, such as silicon or germanium, that has been doped with a pentavalent (five valence electrons) impurity. An N-type material has extra or free electrons in its atomic structure.

nucleus: Core or center of an atom.

O

ohm (Ω): Unit of measurement of electrical resistance. Named in honor of Georg Simon Ohm.

ohmmeter: Meter used to measure resistance in ohms.

Ohm's law: Mathematical relationship between current, voltage, and resistance.

operational amplifier (op-amp): A type of linear integrated circuit capable of mathematical operations (add, subtract, multiply and divide). It is usually used to amplify a tiny signal (like a sine waveform).

OR gate: A type of digital circuit that provides a high logic level at the output when any input is a high logic level.

output device: When referring to a computer system, an output device is a device that is used to communicate to the user (or another computer), the information stored or processed inside the computer. Some examples of output devices are a printer and a monitor.

P

parallel: Method of connecting circuit components side by side, with the ends of each component connected together. Current flowing in the circuit divides among the branches of parallel circuit.

pentavalent: Semiconductor impurity having five valence electrons.

period: Time of one complete cycle of a sine wave.

permeability: Ability of substance to conduct a magnetic field.

photoresisor: Device that changes internal resistance under the influence of light intensity.

photovoltaic cell: Cell that generates voltage at the junction of two materials when exposed to light.

piezoelectric effect: Characteristic of certain crystalline substances to produce a voltage when pressure is applied to its surface.

polarity: Particular state of charged body or terminal, either negative or positive, in respect to another body or terminal.

potential difference: Difference in electrical pressure between two bodies or two points in a circuit. Measured in volts.

primary cell: Cell that cannot be recharged.

primary winding: Coil of a transformer or the coil connected to the power source.

proton: Positively charged particle of an atom found in the nucleus. It has the same mass as the hydrogen nucleus.

P-type material: Semiconductor material, such as silicon or germanium, that has been doped with a trivalent (three valence electrons) impurity. P-type material has a shortage of electrons and has extra, free positive spaces called holes.

R

R: Symbol for resistance in Ohm's law.

random-access memory (RAM): Type of computer memory that temporarily stores information during regular operation of the microprocessor. When the computer is shut off, the information stored in this memory is lost. This type of computer memory is also called the read/write or volatile memory.

RC **time constant:** Time period required for the voltage across a capacitor in an *RC* circuit to increase to 63.2% of maximum value or to decrease to 36.7% of maximum value. Time (in seconds) = *R* (in ohms) multiplied by *C* (in farads).

reactance: Effect of inductance or capacitance in a circuit to the flow of an alternating current. Reactance is calculated in ohms.

read-only memory (ROM): Type of computer memory used to permanantly store information. When the computer is shut off, the information stored in this memory is not lost. Usually ROM provides the information needed for the computer to set itself up after being turned on. Sometimes it is referred to as the bootstrap memory.

rectification: Process of changing alternating current to direct current.

relay: Electromagnetic switch.

reluctance: Resistance to conducting a magnetic field.

residual magnetism: Magnetism remaining in material after magnetizing force is removed.

resistance (R): Quality of an electric circuit that opposes the flow of current. As resistance is overcome, heat is produced.

resonance: Frequency condition in an RLC circuit when the capacitive and inductive reactance are equal and cancel each other out, leaving only resistance in the circuit. Resonant frequency equals:

$$f_r = \frac{1}{2\pi \sqrt{LC}}$$

when, f_r equals the resonant frequency, π equals 3.1416, L equals inductance in henrys, and C equals capacitance in farads.

reverse bias: External voltage applied to a semiconductor PN junction to reduce the flow of current through the junction by widening the depletion region. The opposite of forward bias.

root-mean-square (rms) value: Effective value of alternating current or voltage.

S

saturation: When maximum current flow passes from the emitter to the collector in a transistor. A saturated transistor acts as a closed switch.

schematic: Chart or diagram showing the arrangement and connection of various electronic parts using conventional signs and symbols.

secondary cell: Cell that can be reactivated or recharged by reversing the current through it, which reverses the chemical action.

secondary winding: Coil of a transformer in which voltage is induced by the primary coil. Output is taken from the secondary coil.

self-inductance: Condition in which emf is self-induced in a conductor carrying a current.

semiconductor: Conductor with resistivity in the range between conductors and insulators.

series: Number of circuit components connected in succeeding order or end-to-end, so that current flowing through one will also flow through all of the other components.

series circuit: An electrical circuit with only one path for electrons to flow.

series multiplier: Precision resistor connected in series with a voltmeter to increase its measurement range.

series winding: Type of winding in a motor where the field carries the same current as the armature. (The winding is in series with the armature rather than in parallel with it.)

short circuit: Direct connection across a circuit that provides a zero resistance path for the current flow.

shunt: Component connected across a circuit or in parallel with another component.

shunt resistor: Precision resistor connected in parallel to an ammeter to increase the range of the meter.

shunt winding: Type of winding in a motor in which the field circuit and armature circuit are connected in parallel.

siemens (S): Unit of measurement of conductance. It is the reciprocal of resistance (one divided by the resistance).

single phase: Only one alternating current or voltage produced or used.

sinusoidal (sine) wave: Waveform of a single frequency alternating current.

slip rings: Metal rings connected to a rotating armature of a generator. Brushes contacting these rings pick up the alternating current generated in the armature.

solar: Pertaining to the sun.

solenoid: Coil wound to produce a magnetic field.

south pole: The opposite polarity and end from the north pole of a magnet.

specific gravity: Weight of a liquid or substance as compared to an equal amount of water.

static electricity: Electricity at rest. Considered as a charge of electricity on a body, either negative or positive, compared to moving electricity called current.

step down: To reduce from a higher voltage to a lower voltage, as in a step-down transformer.

step up: To increase to a higher voltage, as in a step-up transformer.

switch: Device for directing or controlling electrical current in a circuit.

T

thermocouple: Device made of two dissimilar metals that are welded together and will produce a small voltage when heated.

transformer: Device that transfers energy from one circuit to another by electromagnetic induction.

transistor: Semiconductor device, usually made of silicon or germanium, having three electrodes. The word *transistor* is derived from two words, *transfer* and *resistor*.

trivalent: Semiconductor impurity having three valence electrons (acceptor impurity).

tune: Process of adjusting a radio receiver so its resonance frequency is the same as that of the transmitting station.

V

valence electron: The electrons that orbit the nucleus in the outermost energy shell or ring of an atom.

vector: Straight line drawn to scale, showing direction and magnitude of a force.

volt (V): Unit of measurement of electrical pressure or electromotive force. One volt will cause one ampere to flow through one ohm of resistance. Named in honor of Alessandro Volta.

voltage (V): Same as electromotive force (emf) and potential difference.

voltage drop: The difference in voltage between two points in a circuit.

voltmeter: Meter used to measure voltage.

W

watt (W): Unit of measurement of electrical power. It is the rate of work accomplished when one ampere of current flows at one volt pressure.

watt-hour (Wh): Unit of electric energy measurement, equal to one watt per hour.

Watt's law: Mathematical relationship between current, voltage, and power.

work (W): Scientifically speaking, work is accomplished when a force acts on a body and moves it. Work is the product of force times distance.

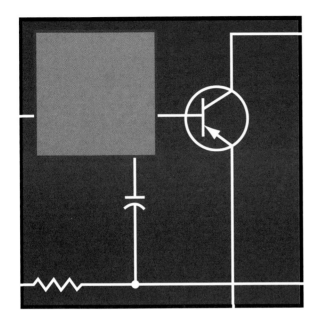

Index

A

alkaline cell, 68
alpha (α), 148
alternating current (ac), 98, 103–109
aluminum, 18
American Wire Gauge (AWG), 13
ammeter, 23, 26
ampere, 16–17
Ampère, André-Marie, 17
ampere-hours, 72
ampere-turns, 81
amplification, 139–140
analog circuit, 155
analog meter, 25, 32
AND gate, 156
armature, 88–89
atom, 10–11
atomic number, 11
atomic weight, 11
autopolarity, 23
autoranging, 23
autotransformer, 137
average value, 108

B

base, 146
basing diagram, 154
battery
    capacity, 72
    cell connections, 66–67
    disposal, 71
    lead-acid storage, 69–71
    primary cell, 67–68
    safety, 70
    secondary cell, 68–71
    series and parallel connections, 71–72
    voltaic cell, 65–66
battery charger, 70
beta (β), 148–149
bleeder resistor, 146
bridge rectifier, 145
brushes, 90

C

capacitance, 111–119
    factors determining, 114–115
capacitive circuit, 112–113
capacitive reactance, 117–118
capacitor, 111–112
    labeling, 114
    testing, 115
    types, 115–117
    working voltage, 113–114
carbon-zinc cell, 67

cell connections, 66–67
    in parallel, 71–72
    in series, 71
ceramic capacitor, 116
circuit breaker, 17, 39
circular mil, 13
clamp-on meter, 24
coil action in a dc circuit, 124
collector, 146
common base circuit, 147
common collector circuit, 147
common emitter circuit, 147
    and beta, 148–149
commutator, 90–91
    dc generator, 98–99
    motor, 89–90
compound generator, 102
compound motor, 91
computers, 158–160
conductance, 18–19, 58
conductance method, 58–59
conductive pathway, 35
conductor, 11–12, 18
control, 35
control devices, 158
conventional current flow, 17
copper, 18
copper loss, 100, 130
core-type transformer, 131
coulomb, 13
Coulomb, Charles, 13
counterelectromotive force (cemf), 122
coupling, 130
current, 11–12
    direction, 96–97
cutoff, 149

D

Dalton, John, 10
d'Arsonval meter, 25–26
d'Arsonval, Arsène (Jacques), 25
DeForest, Lee, 140
diac, 150
dielectric, 114–115
dielectric constant, 114–115
digital circuit, 154, 156–157
digital meter, 25, 32
diode, 141–142
direct current (dc), 99
direct current generator, 95–102
    commutator, 98–99
    output, 99–100
    losses, 100–101
    field excitation, 101–102
doping, 141
driving signal, 140
dry cell, 67–68

E

eddy current loss, 101, 130
effective value, 108–109
electrical circuit, 15–18, 35
electric motor. *See* motor.
electrode, 66
electrolyte, 66
electrolytic capacitor, 116
electromagnet, 81
electromotive force (emf), 16
electron, 11
electron flow, 17
electroscope, 66
electrostatic field, 12, 114
elements, 10
    classification, 11
emitter, 146
energy, 45
equal resistors in parallel, 56–57
equivalent resistance, 59–60

F

farad, 113
Faraday, Michael, 96
field excitation, 101–102
field magnet, 88
field poles, 92
field windings, 91
filter, 145–146
fixed capacitor, 115–116
forward bias, 142

Index

frequency, 106
friction, 73–75
full-wave rectifier, 144–145
fuse, 17, 38–39

G

galvanometer, 96
generator, 97–98. *See also* direct current generator.

H

half-wave rectifier, 144
Hall effect device, 84
heat, 75
henry, 122
Henry, Joseph, 123
hertz, 106
Hertz, Heinrich Rudolph, 106
holes, 140
horsepower, 45–46
hydrometer, 71
hysteresis loss, 100–101, 131

I

impedance, 118–119, 125
inductance, 121–127
induction, 84
induction coil, 135–137
inductive reactance, 124–125
inductor, 121
    labeling, 123–124
input device, 158
insulator, 12
integrated circuit, 153–160
ionization, 11
IR drop, 51
iron-vane meter, 26
isolation transformer, 131–132

K

kilowatt, 43
kilowatt-hour, 43–44
Kirchhoff's current law
    parallel circuit, 57–58
    series circuit, 50

Kirchhoff's voltage law
    parallel circuit, 57–58
    series circuit, 51–52

L

laminating, 101
law of charges, 12–13
lead-acid storage battery, 69–71
left hand rule for a coil, 80, 92
left hand rule for a conductor, 96–97
Lenz, Heinrich Friedrich Emil, 130
Lenz's law, 130
light, 76
light-emitting diode (LED), 142–143
lightning, 75
linear integrated circuit, 154–156
liquid crystal display (LCD), 143
load, 35
local action, 66
lodestone, 77–78
logic gates, 156–157

M

magnet, 78–79
magnetic field, 78
magnetic field pictures, 80
magnetic flux, 78
magnetic shielding, 84–85
magnetism, 77–85
    and electricity, 80–81
    laws, 79
mass number, 11
memory, 159–160
mercury cell, 67–68
meters, 23–33
    ammeters and meter shunts, 26
    analog/digital, 25
    connections, 29, 32
    d'Arsonval, 25–26
    iron-vane, 26
    ohmmeters, 28–29
    overload, 24
    precautions, 32
    reading, 30–31
    voltmeters and multipliers, 27–28

microfarad, 113
microprocessor, 158
mil, 13
milliammeter, 26
molecule, 10
motor, 47, 87–94
    action, 93
    circuits, 90–92
    commutator, 89–90
    operation, 87–89
    principles, 92–93
multimeter, 24, 28

N

NAND gate, 157
neutron, 11
nickel-cadmium (NICAD) cell, 68
NOR gate, 157
NOT gate, 157
NPN transistor, 146
N-type material, 141
nucleus, 11

O

Oersted, Hans Christian, 80
ohm, 17–18
Ohm, Georg Simon, 37
ohmmeter, 23, 28–29
Ohm's law, 27, 35–41
operational amplifier, 155–156
OR gate, 156
oscilloscope, 109
output device, 160
overload, 24
overload protection, 38–39

P

parallel circuit, 55–64
    applications, 60–62
    conductance, 58
    equal resistors in parallel, 56–57
    equivalent resistance, 59–60
    two or more resistors in parallel, 58–59
    unequal resistors in parallel, 57–58
peak-to-peak value, 109
peak value, 108
period, 107
Periodic Table of the Elements, 10
permanent magnet, 78–79
permeability, 78–79
phase displacement, 107
photocell, 76
photoresistor, 76
photovoltaic cell, 76
picofarad, 113
piezoelectric effect, 75
PIRE wheel, 45
PNP transistor, 146
potential difference, 16, 66
potentiometer, 20
power, 43–48
power, conservation, 46–47
power law, 44–45
power transmission, 133–135
powers of 10, 13, 38
pressure, 75
primary cell, 67–68
primary winding, 130
product over the sum method, 57–58
proton, 11
P-type material, 141
pulsating direct current, 99

Q

quality (Q), 123
quartz crystals, 75

R

random-access memory (RAM), 159–160
RC time constants, 117
read-only memory (ROM), 159
reciprocal method, 58–59
rectification, 144–146
reed relay, 82–83
reflecting galvanometer, 25

relay, 81–83
reluctance, 79
residual magnetism, 81
resistance, 12, 17–18
    and conductor size, 18
    uses for, 19–20
resistor color code, 20–21
resonance, 126–127
resonant frequency, 126
reverse bias, 142
rheostat, 20
root-mean-square (rms) value, 108–109

S

saturation, 149
secondary cell, 68–71
secondary winding, 130
self inductance, 121–122
semiconductor, 12, 139–151
    materials, 140–141
semiconductor current, 140–141
series circuit, 49–54
    summary of laws, 52
    voltage drop, 50–52
series generator, 102
series motor, 91
series multiplier, 27
shell-type transformer, 131
short circuit, 39
shunt, 26–27
shunt generator, 102
shunt motor, 91
siemens, 58
Siemens, von, Ernst Werner, 58
silicon controlled rectifier (SCR), 149–150
silver, 18
sine wave, 104–106
    average values, 107–108
    effective values, 108–109
    frequency, 106
    peak-to-peak values, 109
    period, 107
single-phase generator, 107
sinusoidal waveform. *See* sine wave.
slip rings, 104

solar cell, 76
solenoid, 81
    as a switch, 83–84
sources of electricity, 65
specific gravity, 71
static electricity, 10, 73–75
static machine, 74
step-down transformer, 132
step-up transformer, 132
stranded conductor, 19
superconductor, 19
switches, 39–41

T

tantalum capacitor, 116–117
Tesla, Nikola, 89
thermocouple, 75
three-phase generator, 107
thyristor, 149
time constant, 117, 124
tolerance, 21
transformer, 129–137
    autotransformer, 137
    construction, 131
    induction coil, 135–137
    losses, 130–131
    power transmission, 133–135
    turns ratio, 131–132
transistor, 146
    invention of, 147
transistor amplifier, 147–148
transistor circuits, 146–147
transistor switch, 149
triac, 150
turns ratio, 131–132
two-phase generator, 107

U

unijunction transistor (UJT), 150–151
unity coupling, 130

V

valence electrons, 140
Van de Graaff generator, 74

variable capacitor, 115
variable resistor, 20
vector, 104
volt, 15–16
Volta, Alessandro, 66
voltage, 16
    from friction, 73–75
    from heat, 75
    from light, 76
    from pressure, 75
voltage drop, 50–52
voltage source, 35
voltaic cell, 65–66
voltmeter, 23, 27, 29

W

watt, 43
watt-hour, 43
Watt, James, 46
Watt's law, 44–45
wire sizes, 13–14
work, 45
working voltage, 113–114

X

X_L and X_C, combining in a circuit, 125–126